Science
(AQA)

Complete Revision and Practice

Nigel Saunders

Published by BBC Active, an imprint of Educational Publishers LLP, part of the Pearson Education Group Edinburgh Gate, Harlow, Essex CN20 2JE, England

Text copyright © Nigel Saunders 2008
Design & concept copyright © BBC Active 2008, 2010

BBC logo © BBC 1996. BBC and BBC Active are trademarks of the British Broadcasting Corporation

First published 2008
This edition 2010
10 9 8 7 6 5 4 3

ISBN 978-1-4066-5447-9

Printed in China CTPSC/03

Minimum recommended system requirements
PC: Windows(r), XP sp2, Pentium 4 1 GHz processor (2 GHz for Vista), 512 MB of RAM (1 GB for Windows Vista), 1 GB of free hard disk space, CD-ROM drive 16x, 16 bit colour monitor set at 1024 x 768 pixels resolution
MAC: Mac OS X 10.3.9 or higher, G4 processor at 1 GHz or faster, 512 MB RAM, 1 GB free space (or 10% of drive capacity, whichever is higher), Microsoft Internet Explorer® 6.1 SP2 or Macintosh Safari™ 1.3, Adobe Flash® Player 9 or higher, Adobe Reader® 7 or higher, Headphones recommended

If you experiencing difficulty in launching the enclosed CD-ROM, or in accessing content, please review the following notes:
1 Ensure your computer meets the minimum requirements. Faster machines will improve performance.
2 If the CD does not automatically open, Windows users should open 'My Computer', double-click on the CD icon, then the file named 'launcher.exe'. Macintosh users should double-click on the CD icon, then 'launcher.osx'
Please note: the eDesktop Revision Planner is provided as-is and cannot be supported.
For other technical support, visit the following address for articles which may help resolve your issues:
http://centraal.uk.knowledgebox.com/kbase/

If you cannot find information which helps you to resolve your particular issue, please email: Digital.Support@pearson.com.
Please include the following information in your mail:
- Your name and daytime telephone number.
- ISBN of the product (found on the packaging.)
- Details of the problem you are experiencing - e.g. ho................ges etc.
- Details of your computer (operating system, RAM, pr................

Contents

Biology

Unit B1a Human Biology

Unit B1b Evolution and Environment

Chemistry

Unit C1a Products from Rocks

Physics

* Only available in the CD-ROM version of the book

Exam board specification map

These specifications have identical content, covering the whole programme of study for KS4 Science, although the assessment styles for Science A and Science B are different.

Science A The specific feature of this specification is that external assessment is available through 'bite-size' objective tests. Each of the three units, Biology 1, Chemistry 1 and Physics 1, is divided into two equal sections and each section is examined in a separate 30 minute test. The tests are available in November, March and June. The objective tests are available as paper-based and on-screen tests in centres.

Science B In contrast, Science B does not offer assessment through the 'bite-size' test route but has 45-minute written papers with structured questions. There is one paper for each of Biology 1, Chemistry 1 and Physics 1, available in January and June.

Topics	AQA Science A	AQA Science B
Biology		
Unit B1b Evolution and Environment		
The nervous system	✓	✓
Hormones	✓	✓
Controlling reproduction	✓	✓
Diet and exercise	✓	✓
Cholesterol and salt	✓	✓
Medical drugs	✓	✓
Recreational drugs	✓	✓
Pathogens and disease	✓	✓
Antibiotics	✓	✓
Vaccination	✓	✓
Unit B1b Evolution and Environment		
Adaptation	✓	✓
Reproduction	✓	✓
Cloning and genetic engineering	✓	✓
Extinction and evolution	✓	✓
Natural selection	✓	✓
Pollution	✓	✓
Global warming	✓	✓
Chemistry		
Unit C1a Products from Rocks		
Atoms and elements	✓	✓
Reactions and compounds	✓	✓
Limestone	✓	✓
Ores and iron	✓	✓
Alloys	✓	✓

Topics	AQA Science A	AQA Science B
Copper, titanium and aluminium	✓	✓
Hydrocarbons	✓	✓
Burning fuels	✓	✓
Unit C1b Oils, Earth and Atmosphere		
Alkenes	✓	✓
Polymers	✓	✓
Plant oils	✓	✓
Food additives	✓	✓
Structure of the Earth	✓	✓
The atmosphere	✓	✓
Noble gases	✓	✓
Physics		
Unit P1a Energy and Electricity		
Heat transfer	✓	✓
Efficiency	✓	✓
Cost of electricity	✓	✓
The National Grid	✓	✓
Generating electricity	✓	✓
Renewable energy sources	✓	✓
Resources compared	✓	✓
Unit P1b Radiation and the Universe		
The electromagnetic spectrum	✓	✓
Some uses of electromagnetic radiation	✓	✓
Communication signals	✓	✓
Hazards of electromagnetic radiation	✓	✓
Atomic radiation	✓	✓
Half-life	✓	✓
Observing the universe	✓	✓

Introduction

How to use GCSE Bitesize Complete Revision and Practice

Begin with the CD-ROM. There are five easy steps to using the CD-ROM – and to creating your own personal revision programme. Follow these steps and you'll be fully prepared for the exam without wasting time on areas you already know.

Topic checker

Step 1: Check

The Topic checker will help you figure out what you know – and what you need to revise.

Revision planner

Step 2: Plan

When you know which topics you need to revise, enter them into the handy Revision planner. You'll get a daily reminder to make sure you're on track.

Step 3: Revise

From the Topic checker, you can go straight to the topic pages that contain all the facts you need to know.

- Give yourself the edge with the Web*Bite* buttons. These link directly to the relevant section on the BBC Bitesize Revision website.

- Audio*Bite* buttons let you listen to more about the topic to boost your knowledge even further. *

Step 4: Practise

Check your understanding by answering the Practice questions. Click on each question to see the correct answer.

Exam Bite

Step 5: Exam

Are you ready for the exam? Exam*Bite* buttons take you to an exam question on the topics you've just revised. *

* Not all subjects contain these features, depending on their exam requirements.

Interactive book — You can choose to go through every topic from beginning to end by clicking on the Interactive book and selecting topics on the Contents page.

Exam questions — Find all of the exam questions in one place by clicking on the Exam questions tab.

Last-minute learner — The Last-minute learner gives you the most important facts in a few pages for that final revision session.

You can access the information on these pages at any time from the link on the Topic checker or by clicking on the Help button. You can also do the Tutorial which provides step-by-step instructions on how to use the CD-ROM and gives you an overview of all the features available. You can find the Tutorial on the Home page when you click on the Home button.

Other features include:

Click on the draw tool to annotate pages. N.B. Annotations cannot be saved.

 **Page turn**

Click on Page turn to stop the pages turning over like a book.

Click on the Single page icon to see a single page.

Click on this arrow to go back to the previous screen.

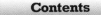 **Contents**

Click on Contents while in the Interactive book to see a contents list in a pop-up window.

Click on these arrows to go backward or forward one page at a time.

Click on this bar to switch the buttons to the opposite side of the screen.

Click on any section of the text on a topic page to zoom in for a closer look.

N.B. You may come across some exercises that you can't do on-screen, such as circling or underlining, in these cases you should use the printed book.

About this book

Use this book whenever you prefer to work away from your computer.
It consists of two main parts:

 A set of double-page spreads, covering the essential topics for revision from each area of the curriculum. Each topic is organised in the following way:

- a summary of the main points and an introduction to the topic

- lettered section boxes cover the important areas within each topic

- key facts highlighting essential information in a section or providing tips on answering exam questions

- practice questions at the end of each topic to check your understanding.

 A number of special sections to help you consolidate your revision and get a feel for how exam questions are structured and marked. These extra sections will help you to check your progress and be confident that you know your stuff. They include:

- Topic checker – quick questions covering all topic areas

- exam-style questions and worked model answers and comments to help you get full marks

- Complete the facts – check that you have the most important ideas at your fingertips

- Last-minute learner – the most important facts in just a few pages.

About your exam

Get organised

You need to know when your exams are before you make your revision plan. Check the dates, times and locations of your exams with your teacher, tutor or school office.

On the day

Aim to arrive in plenty of time, with everything you need: several pens, pencils and a ruler.

On your way, or while you're waiting, read through your Last-minute learner.

In the exam room

When you are issued with your exam paper, you must not open it immediately. However, there are some details on the front cover that you can fill in before you start the exam itself (your name, centre number, etc.). If you're not sure where to write these details, ask one of the invigilators (teachers supervising the exam).

When it's time to begin writing, read each question carefully. Remember to keep an eye on the time.

Finally, don't panic! If you have followed your teacher's advice and the suggestions in this book, you will be well prepared for any question in your exam.

Picking up marks in exams

The examiners are not trying to trick you. They just want to give you the chance to show what you can do. But they are not mind readers. They can only mark what you actually do on the examination paper. Here are some general tips:

- Read through the whole paper for a few minutes at the start and tick the easiest questions as you go.

- Answer your ticked questions first. This will help you feel more calm and confident.

- Tackle the remaining questions.

- Remember to read the questions carefully and underline important words. The information at the start is there to help you, so don't ignore it.

More advice for written exams

Keeping to time

- The written exams work out at one mark per minute on average. You are likely to run out of time if you spend ten minutes on a question worth just one mark.

- The amount of space and the number of marks available are clues to how much to write. You probably won't have written enough if you write one line when there are four marks and six lines.

Calculations

- Write down the equation you need to answer the question.

- Show your working out.

- Show the units if asked.

- Check that the answer makes sense.
  For example, is a snail really likely to move at 100 m/s on its own?

Graphs

- Use a sharp pencil.

- You are likely be given a partly completed graph to finish in the Foundation Tier exam.

In the Higher Tier exam you are likely to have to draw your own axes. Try to use more than half the area of the graph paper without choosing a strange scale.

- Label both axes and include the units.

- Plot each point with a neat cross, or use a ruler to draw bars.

- In a bar chart, don't waste time shading in the bars unless the question asks you to. In a line graph, draw a line or smooth curve with a single stroke of your pencil.

The nervous system

- Receptors detect changes in the environment called stimuli.
- Information from receptors travels along neurones to the brain.
- The brain coordinates the response.
- Reflex actions are fast and automatic.

A Receptors and stimuli

1 A stimulus is a change in the surroundings. Stimuli include sound, light, chemicals, touch, pain, pressure, temperature and changes in position.

2 **key fact** **A receptor is a cell or group of cells that can detect a stimulus. For example, receptors in the ear detect sound, receptors in the eye detect light, and receptors in the tongue detect chemicals in food and drink.**

> **exam tip >>**
> Ears, eyes and the tongue are sense organs. You don't need to know about their structure or function for the exams.

B Neurones and the brain

1 Nerve cells are called neurones. They carry information from one place to another in the body as electrical signals.

2 **key fact** **Information travels as electrical impulses along nerve cells from the receptors to the brain.**

3 The brain coordinates the response to the stimulus. The response involves effectors and other nerve cells.

C Synapses

1 **key fact** **A synapse is the gap between two nerve cells.**

2 Electrical impulses cannot cross the synapse. Instead, chemicals called neurotransmitters diffuse across it. They are produced by one of the nerve cells, and make the other nerve cell send information.

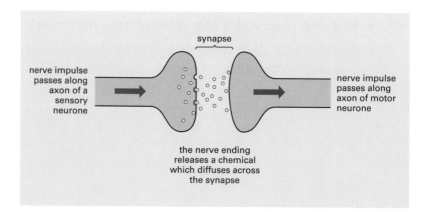

nerve impulse passes along axon of a sensory neurone

synapse

nerve impulse passes along axon of motor neurone

the nerve ending releases a chemical which diffuses across the synapse

D Reflex actions

❶ **key fact** Reflex actions are fast and automatic. They do not need to be learned and they happen without us thinking about them. They involve receptors, different types of nerve cell, and effectors.

❷ Sensory neurones carry information from receptors.

Relay neurones in the spinal cord carry information from sensory neurones to motor neurones.

Motor neurones carry information to effectors.

Effectors are muscles and glands.

❸ **key fact** The sequence of events is:

stimulus
↓
receptor
↓
sensory neurone
↓
relay neurone
↓
motor neurone
↓
effector

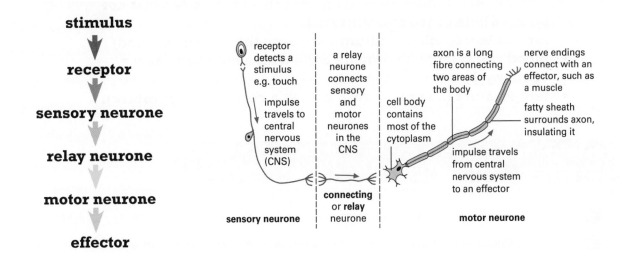

receptor detects a stimulus e.g. touch

impulse travels to central nervous system (CNS)

a relay neurone connects sensory and motor neurones in the CNS

cell body contains most of the cytoplasm

axon is a long fibre connecting two areas of the body

nerve endings connect with an effector, such as a muscle

fatty sheath surrounds axon, insulating it

impulse travels from central nervous system to an effector

sensory neurone

connecting or **relay** neurone

motor neurone

>> practice questions

1 **What is a synapse?**

2 **You rapidly pull your hand away if you accidentally touch something hot. Outline the sequence of events in this reflex action.**

Hormones

- The body's internal conditions are controlled.
- The nervous system and hormones are involved in this control.
- Hormones are chemicals secreted by glands.
- The bloodstream transports hormones to their target organs.

A Some conditions to control

1. The body's cells and the chemicals they contain are damaged if the conditions change too much.

2. **key fact** The internal conditions that must be controlled within tight limits include:
   - water content (the different chemicals in the body must not be too dilute or too concentrated)
   - ion content (such as sodium ions from the salt in our food)
   - blood sugar level (to provide cells with a constant source of energy)
   - temperature (enzymes work best at a certain temperature).

B Controlling water and ion content

1. Water leaves the body from the lungs when we breathe out.
2. Water and ions leave the body from the skin when we sweat and from the kidneys in urine.

C Sports drinks

The manufacturers of sports drinks claim that their products are better than ordinary drinks when people take exercise. Sports drinks contain dissolved ions and sugars. The aim is to replace the water, ions and sugars lost from the body during exercise.

exam tip >>
You may be given information in a table or graph about sports drinks to evaluate in the exam.

4

D Hormones

1 **key fact** Hormones are chemicals produced or secreted by glands. They are transported in the bloodstream to their target organs, where they cause a response or change.

2 Many processes in the body are coordinated by hormones. For example:

- ADH controls the amount of urine produced by the kidneys
- insulin makes the liver convert glucose to glycogen
- adrenaline makes the heart beat faster
- sex hormones control puberty and reproduction.

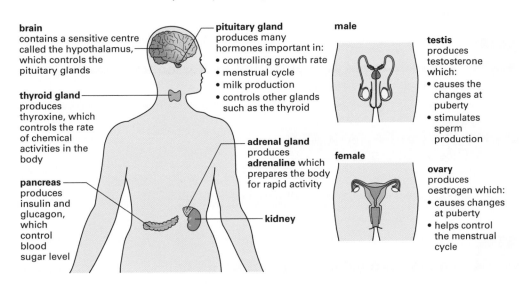

brain
contains a sensitive centre called the hypothalamus, which controls the pituitary glands

thyroid gland
produces thyroxine, which controls the rate of chemical activities in the body

pancreas
produces insulin and glucagon, which control blood sugar level

pituitary gland
produces many hormones important in:
- controlling growth rate
- menstrual cycle
- milk production
- controls other glands such as the thyroid

adrenal gland
produces **adrenaline** which prepares the body for rapid activity

kidney

male

testis
produces testosterone which:
- causes the changes at puberty
- stimulates sperm production

female

ovary
produces oestrogen which:
- causes changes at puberty
- helps control the menstrual cycle

E Nervous system and hormonal system compared

nervous system	hormonal system
information travels very quickly	hormones are transported more slowly
response is short-lived	response is often long-lasting
usually act on a small area	often have widespread effects

>> practice questions

1 Name two internal conditions that must be controlled by the body.

2 How are hormones transported in the body?

3 Suggest an advantage of being able to coordinate life processes by two systems, one involving nerves and the other involving hormones.

Controlling reproduction

- The menstrual cycle is controlled by several hormones.
- Hormones can be used to control fertility in women.

A The menstrual cycle

1 Having a period is called menstruation. A woman's monthly cycle is called the menstrual cycle. It lasts about 28 days.

2 **key fact** During the menstrual cycle, an egg is released from an ovary and there are changes in the thickness of the lining of the womb.

3 **key fact** These changes are controlled by hormones secreted by the pituitary gland and ovaries.

B FSH, oestrogen and LH

1 FSH, oestrogen and LH are three hormones involved in controlling the menstrual cycle.

2 **key fact** FSH is secreted by the pituitary gland:

- it causes an egg to mature in an ovary
- it stimulates the ovaries to release oestrogen.

3 **key fact** Oestrogen is secreted by the ovaries:

- it stops FSH being produced (so that only one egg matures in a cycle)
- it stimulates the pituitary gland to release LH
- it causes the lining of the womb to thicken.

4 LH is secreted by the pituitary gland. It causes an egg to be released at around day 14.

Progesterone is a hormone that maintains the lining of the womb if the egg is fertilised.

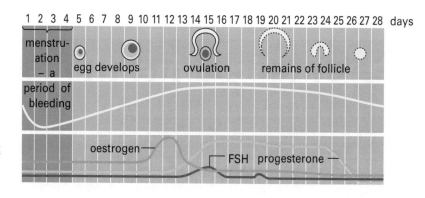

menstruation – a period of bleeding | egg develops | ovulation | remains of follicle

oestrogen — FSH progesterone —

6

C Oral contraceptives

1 key fact **Oral contraceptives prevent women becoming pregnant. They contain hormones that prevent FSH being produced. This means that eggs do not mature in the ovaries.**

2 Oral contraceptives allow couples to choose when to start a family.

3 Oral contraceptives cause problems for some women, including weight gain and mood swings.

D Fertility drugs

1 key fact **One reason why a woman might be infertile is that her level of FSH is too low to cause eggs to mature. Fertility drugs containing FSH increase the chance of a woman becoming pregnant. They boost the level of FSH and cause eggs to mature.**

2 Fertility drugs increase the chances of couples starting a family if they are having difficulty doing so.

E IVF

1 IVF stands for *in vitro* fertilisation. It involves fertilising a woman's egg outside her body in the laboratory. The embryo is allowed to develop for a short time and then implanted into the woman's womb.

2 IVF can help infertile couples to have a family. The embryo can be tested for genetic disorders. But some people worry that couples may want 'designer babies'. For example, they may want to have a boy rather than a girl, or an embryo with a particular feature, such as high intelligence.

>> practice questions

1 Where is FSH secreted from, and what does it do?

2 Where is oestrogen secreted from, and what does it do?

3 Explain how FSH and oestrogen may be used to control fertility in women.

Diet and exercise

A person can become too thin or too fat if their diet is unbalanced.

Some people in the developing world suffer illness owing to lack of food.

Some people in the developed world suffer illness owing to too much food and too little exercise.

A A balanced diet

① key fact A balanced diet contains the correct nutrients in the right amounts. It supplies the right amount of energy for the amount of activity a person does.

The table below shows the main nutrients, why they are needed, and some typical sources of them. Fibre and water are also needed in the diet.

nutrient	why we need it	typical source
carbohydrate	energy	potatoes, rice
protein	growth and repair	eggs, meat
fat	energy	butter, vegetable oils
vitamins	maintaining health	fruit, vegetables, bread
minerals	maintaining health	meat, vegetables

② Malnutrition happens if a diet is not balanced. The effects of malnourishment include:

* becoming too thin or too fat
* deficiency diseases if the amount of a nutrient in the diet is too low
* diseases linked to excess weight if the amount of energy supplied by the diet is too high.

B Deficiency diseases

>> key fact Some people in the developing world can suffer from a lack of sufficient food. They suffer health problems, including irregular periods in women and a reduced resistance to infection.

C Exercise and diet

① **key fact** The rate at which the various chemical reactions take place in your body's cells is called the metabolic rate. It is affected by various factors, including:

- the amount of muscle you have compared to your body fat
- the amount of activity you do
- inherited factors.

② Exercise increases the metabolic rate, and it stays higher for some time after the exercise is over.

D Excess weight

① **key fact** Less food is needed when the surroundings are warm, and when a person is taking little exercise. For example, someone sitting at a desk in an office needs less energy from their food than someone working outside in a physically demanding job.

② If someone eats too much food, and takes too little exercise, they become overweight. Very overweight people are described as obese. High levels of obesity are a problem in the developed world.

③ **key fact** Diseases linked to excess weight include:

- diabetes (high levels of sugar in the blood)
- heart disease
- arthritis (worn and swollen joints)
- high blood pressure.

>> practice questions

1 What is a balanced diet?

2 Outline some problems associated with too little food in the diet.

3 Outline some problems associated with too much food in the diet.

Cholesterol and salt

- High levels of cholesterol in the blood increase the risk of heart disease.
- LDLs are 'bad' cholesterol and HDLs are 'good' cholesterol.
- Unsaturated fats in the diet help to improve the balance between LDLs and HDLs.
- Excess salt in the diet can cause high blood pressure.

A Cholesterol

1 key fact Cholesterol is a substance found in the blood. It is made by the liver from saturated fats in the diet.

2 key fact The amount of cholesterol produced by the liver depends upon inherited factors and diet:

- saturated fats increase blood cholesterol levels
- unsaturated fats help to decrease blood cholesterol levels.

3 key fact High levels of cholesterol in the blood increase the chance of developing cardiovascular disease – heart disease and blocked arteries.

Arteries have thick, muscular, elastic walls. Fatty deposits can narrow arteries and may block them.

B LDLs and HDLs

1 Cholesterol is transported around the body in the bloodstream attached to proteins. The combination of cholesterol and protein is called lipoprotein.

2 key fact There are two types of lipoprotein:

- low-density lipoprotein or LDL
- high-density lipoprotein or HDL.

3 key fact LDLs carry cholesterol from the liver to the cells of the body. They are sometimes called 'bad' cholesterol.

④ **key fact** HDLs carry excess cholesterol from the cells back to the liver. They are sometimes called 'good' cholesterol.

C   Lipoproteins and health

① **key fact** About 30% of blood cholesterol is carried as HDLs. The balance of LDL to HDL is important for health:

- higher proportions of LDLs increase the risk of cardiovascular disease

- higher proportions of HDLs reduce the risk cardiovascular disease.

② Saturated fats increase blood cholesterol levels. Foods such as meat and eggs are rich in saturated fats.

③ Mono-unsaturated fats and polyunsaturated fats may help to reduce cholesterol levels in the blood, and improve the balance of LDL to HDL. Foods such as olive oil and oily fish are rich in unsaturated fats.

exam tip   >>

Statins are a type of medicine that reduces blood cholesterol levels. You may be given information in the exam so you can evaluate the effect of statins on cardiovascular disease.

D   Processed foods

① **key fact** Processed foods often contain a lot of fat. Too much of these foods in the diet can lead to obesity and raised blood cholesterol levels. This in turn increases the risk of cardiovascular disease.

② **key fact** Processed foods often also contain a lot of salt. Too much salt causes increased blood pressure in about 30% of the population.

③ High blood pressure increases the risk of cardiovascular disease.

>> practice questions

1 Where is cholesterol made in the body?

2 Outline some differences between LDLs and HDLs.

3 Explain how eating too much processed food can lead to ill health.

Medical drugs

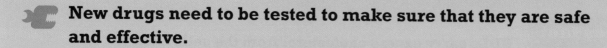

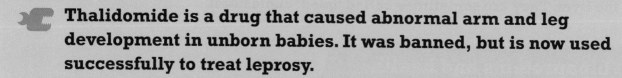

- New drugs need to be tested to make sure that they are safe and effective.

- Thalidomide is a drug that caused abnormal arm and leg development in unborn babies. It was banned, but is now used successfully to treat leprosy.

A Medical drugs

1 **key fact** Drugs are substances that change chemical processes in the body. These changes can be helpful or harmful, depending on the drug.

2 **key fact** Many drugs are found naturally. Such drugs have been known about for many years.

3 For example, the ancient Greeks used willow bark to help cure fevers and pains. Salicylic acid is the active ingredient in willow bark. It has a similar chemical structure to aspirin. Natural substances are a source of raw materials for developing new drugs.

salicyclic acid   aspirin

B Testing drugs

1 **key fact** Scientists are developing new drugs and these must be tested. It is important that new drugs are effective (they actually work) and are safe (they will not cause dangerous side effects).

2 There are three main stages in testing new drugs:

- using computer models and human cells grown in the laboratory. Many substances fail this test because they are toxic and damage cells.

remember >>

It is a legal requirement in the UK that all new drugs are tested on animals before reaching the human trial stage. But it is illegal to test cosmetics and tobacco products on animals.

- new drugs passing the first stage are tested on animals. A known amount is given to the animals, which are then monitored carefully for any side-effects.

- drugs that have passed the second stage are used in clinical trials. They are trialled on healthy volunteers to discover any side effects. They are then tested on people with the illness to ensure they work.

C Thalidomide

1 **key fact** **Thalidomide is a drug that was originally developed as a sleeping tablet in the 1950s.**

2 **key fact** **It was also found to be effective at reducing the symptoms of morning sickness in pregnant women. But it had not been tested for this different use. Between the late 1950s and early 1960s thousands of babies were born with deformities, such as abnormally short arms or legs.**

3 Thalidomide was banned because of its effects on unborn babies. It has been licensed recently to treat leprosy, a disease of the skin and nerves caused by a bacterium.

remember >>

The thalidomide story shows what can happen if drugs are not tested sufficiently before use.
Its effectiveness against morning sickness and leprosy were discovered by accident.

>> practice questions

1 What is a drug?

2 Describe the main stages in testing a new medical drug.

3 Explain why female leprosy sufferers cannot be prescribed thalidomide if they are pregnant or not using contraceptives.

Recreational drugs

- Some people use recreational drugs to alter their mood. Drugs can be legal or illegal.

- People may become addicted to a drug and experience withdrawal symptoms if they stop taking it.

- Tobacco and alcohol are legal recreational drugs. They can damage the body.

A Tobacco

1 Tobacco smoke contains around 4000 different chemicals. Many of these are harmful.

2 **key fact** Nicotine is the addictive substance in tobacco smoke. It creates a powerful dependency so that smokers quickly become addicted. Smokers have unpleasant withdrawal symptoms if they try to give up smoking.

3 **key fact** Tobacco smoke contains carbon monoxide. This reduces the ability of the blood to carry oxygen because it combines with haemoglobin in red blood cells. Smoking during pregnancy is very dangerous. It reduces the amount of oxygen available to the growing fetus. This can cause babies to have a low weight. It can also cause miscarriage and premature birth.

4 The effects of carbon monoxide put extra strain on the circulatory system, and cause an increased risk of cardiovascular disease and strokes.

5 **key fact** Tobacco smoke also contains carcinogens, such as tar. These are substances that cause cancer. Smoking increases the risk of lung cancer, mouth cancer and throat cancer.

remember >>

Carbon monoxide reduces the oxygen-carrying capacity of the red blood cells.

exam tip >>

You may be given information to evaluate in the exam about smoking and health. Use this with your knowledge and understanding of science to answer the question.

B Alcohol

1 Ethanol is the alcohol found in alcoholic drinks such as beer, wine and spirits. It is usually just called alcohol.

2 **key fact** **Alcohol affects the nervous system. It slows down reactions and helps people to relax.**

3 **key fact** **Too much alcohol leads to a lack of self-control. Drinkers may not realise how much they are drinking. They may become unconscious, and may even fall into a coma. Alcohol damages the liver and brain in the longer term.**

C Legal and illegal drugs

1 Alcohol and tobacco are legal recreational drugs.

2 Illegal drugs include substances that are banned by law, and prescription drugs that have been modified. Heroin and cocaine are illegal recreational drugs. They are very addictive, and may cause permanent mental problems.

3 **key fact** **More people use legal recreational drugs than use illegal drugs. So the overall impact on health is greater from legal than illegal drugs.**

4 Drugs change chemical processes in the body, so they all have the potential to damage health. Addictive recreational drugs may damage health indirectly. For example:

- users may turn to crime to pay for their drugs, which affects the lives of other people
- used needles can transmit diseases such as AIDS and hepatitis from infected blood.

>> practice questions

1 What does 'addicted' mean?

2 Explain the problems caused by nicotine, carbon monoxide and tar in tobacco smoke.

3 Which parts of the body are affected by alcohol?

4 Suggest why smokers may be reluctant to give up smoking, even though the link between smoking and increased risk of lung cancer has been firmly established.

Pathogens and disease

- ✦ Pathogens are microorganisms which cause infectious diseases.

- ✦ Bacteria and viruses are pathogens. They reproduce rapidly inside the body and release toxins that make us feel ill.

- ✦ White blood cells ingest pathogens, produce antitoxins, and produce antibodies that destroy particular pathogens.

A Pathogens

1 **key fact** Pathogens are microorganisms that cause infectious diseases. Bacteria and viruses are pathogens.

2 Bacteria are tiny living cells – the largest are only 10 micrometres long (10 μm or 10 thousandths of a millimetre).

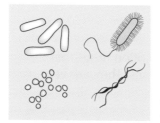

Diseases caused by bacteria include food poisoning, cholera, typhoid and whooping cough.

3 **key fact** Bacteria can multiply rapidly once inside the body. They release poisons or toxins that make us feel ill.

4 Viruses are even smaller than bacteria. They consist of a fragment of genetic material inside a protective protein coat.

5 **key fact** Viruses can only reproduce inside host cells, and they damage the cell when they do this.

6 Once a virus gets inside a cell, it multiplies rapidly. Viruses eventually fill the whole cell, bursting it open and letting the viruses escape.

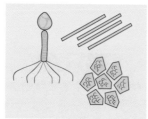

Diseases caused by viruses include colds, influenza (flu), measles, mumps, rubella, tetanus, chicken pox and AIDS.

B White blood cells

1 **key fact** White blood cells help to defend the body against pathogens:

- they produce antitoxins. These are chemicals that counteract the toxins produced by pathogens

- they ingest pathogens.

2 Some of the substances in pathogens and on their surfaces are recognised by the body as foreign. These substances are called antigens. Different pathogens have different antigens.

3 **key fact** White blood cells can produce proteins called antibodies. These can attach to antigens. Different antibodies recognise different antigens. Chemical signals are released when antibodies attach to a pathogen. Two things can happen as a result:

- chemicals are released that destroy the pathogen

- white blood cells are attracted and ingest the pathogen.

remember >>

White blood cells do not eat pathogens. They surround them, engulfing them and destroying them with reactive chemicals.

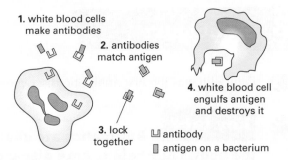

1. white blood cells make antibodies
2. antibodies match antigen
3. lock together
4. white blood cell engulfs antigen and destroys it

⊔ antibody
▯ antigen on a bacterium

exam tip >>

Make sure you show you understand that the pathogens are not the disease. Instead, they are the cause of the disease.

>> practice questions

1 What is a pathogen?

2 How do pathogens make us feel ill?

3 Outline three ways in which white blood cells can help to defend the body against pathogens.

Antibiotics

- Antibiotics kill bacteria but not viruses.
- It is difficult to develop drugs that kill viruses without damaging the body.
- Many strains of bacteria have developed resistance to antibiotics because of natural selection.

A Relieving symptoms

1 key fact Some medicines relieve the symptoms of a disease without actually killing the pathogens.

2 Painkillers are like this. They reduce the feeling of pain, but do not kill the pathogens or repair the damage caused by the pathogens.

B Antibiotics

1 Penicillin was the first antibiotic discovered.

2 key fact Antibiotics are chemicals that kill bacteria inside the body. They help to cure diseases caused by bacteria.

3 key fact Antibiotics cannot kill viruses. Viruses reproduce inside the body's cells. It is difficult to develop drugs that kill viruses without also damaging the body's cells.

4 Antibiotics are produced naturally by certain fungi, but modern antibiotics are designed by scientists and manufactured.

The table below shows four different antibiotics and how they work.

antibiotic	how it works
Penicillin	breaks down cell walls
Erythromycin Vancomycin	stops protein synthesis
Ciprofloxacin	stops DNA replication

C Antibiotic-resistant bacteria

1 Many strains of bacteria have developed resistance to one or more antibiotics. This means that the antibiotics have no effect on them.

2 **key fact** MRSA, or 'methicillin-resistant *Staphylococcus aureus*', is a strain of bacteria that is resistant to antibiotics based on penicillin. It is sometimes called a 'superbug' because of this.

3 **key fact** Bacteria develop antibiotic resistance because of natural selection.

4 In any population of bacteria, some bacteria may naturally be resistant to an antibiotic. When a patient is given this antibiotic, the bacteria are all killed, except for the resistant ones. These are able to reproduce and pass on their resistance to later generations of bacteria.

D Treating infections

1 Ignaz Semmelweis was a hospital doctor in the 19th century. He realised the importance of cleanliness in hospitals. He insisted on keeping the hospital clean and on personal hygiene amongst the staff. This reduced the risk of infection. But his ideas were ignored at the time, because no-one knew that infectious diseases were caused by pathogens that could be killed.

2 When antibiotics were discovered, they were relied upon too much to treat disease and prevent infections. In the recent past, hospitals were not always clean enough to reduce the risk of infection. Nowadays great care is taken to maintain hygiene.

3 **key fact** It is important to avoid over-use of antibiotics. This reduces the chance of new antibiotic-resistant strains developing.

>> practice questions

1 **What is an antibiotic?**

2 **Explain why colds cannot be cured by antibiotics.**

3 **Explain why it is important not to use antibiotics when they are not really needed.**

Vaccination

- People can be immunised against a particular pathogen by vaccination.

- A vaccine is a small amount of a dead or inactive form of a pathogen.

- A vaccine stimulates white blood cells to make antibodies against the pathogen. This makes the person immune to the pathogen.

A Immunity

1. When someone is infected by a particular pathogen, white blood cells are made that can recognise and destroy that pathogen.

2. After the pathogen has been destroyed, some of these white blood cells remain. If the person is reinfected by the same pathogen, the white blood cells reproduce rapidly. They produce a lot of antibodies against the pathogen very quickly. In this way, people can become immune to a particular pathogen.

3. Vaccination allows people to become immune to a particular pathogen without being infected first.

B Vaccination

1. **key fact** **A vaccine consists of small amounts of:**

   - **the dead pathogen, or**

   - **inactive forms of the pathogen, or**

   - **antigens or toxins extracted from the pathogen.**

2. **key fact** **The vaccine is put into the body, usually by injection. It stimulates the production of white blood cells that can make antibodies against a particular pathogen.**

3. Some of these white blood cells remain in the body after the vaccine has been destroyed by the body.

4 **key fact** If the pathogen enters the body in the future, these white blood cells reproduce rapidly. They produce a lot of antibodies against the pathogen very quickly. In this way, the person is immune to the pathogen without having been infected first.

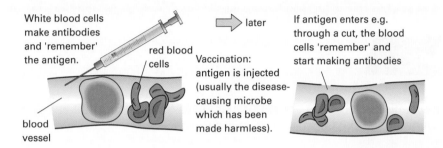

White blood cells make antibodies and 'remember' the antigen.

red blood cells

blood vessel

Vaccination: antigen is injected (usually the disease-causing microbe which has been made harmless).

later

If antigen enters e.g. through a cut, the blood cells 'remember' and start making antibodies

C Some vaccines

1 **key fact** Different vaccines are needed to give protection against different diseases. The MMR vaccination is given to children. It is a mixture of vaccines that protects them against measles, mumps and rubella (German measles).

2 Vaccines against diphtheria, tetanus, pertussis (whooping cough) and polio are also commonly given.

3 **key fact** Pathogens can change or mutate in such a way that a vaccine may no longer be effective. This is particularly true of the influenza (flu) virus. Slightly different vaccines are needed each year.

>> practice questions

1 What is a vaccine?

2 Explain how vaccination works.

3 Elderly people are particularly at risk from the symptoms of flu.
   Suggest why it is recommended that they receive a flu vaccination every autumn.

Adaptation

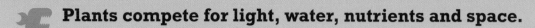 Plants compete for light, water, nutrients and space.

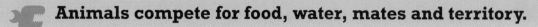 Animals compete for food, water, mates and territory.

➤ Organisms have adaptations that allow them to survive in their environment.

A Competition

① The place where an organism such as an animal or plant lives is called its habitat. The conditions surrounding the organism are the environment.

② **key fact** Organisms need to have a supply of materials from their surroundings to survive. They may also need materials from other organisms. For example, animals and plants need water, and animals must eat plants or other animals.

③ Organisms can only survive if they can get enough resources. A given habitat may only have a limited amount of a particular resource. Organisms must then compete with other organisms for this resource. If they are unsuccessful and cannot move to another habitat, they will die.

④ **key fact** Plants may compete with each other for:

- light

- water

- space

- nutrients from the soil, such as mineral salts.

⑤ **key fact** Animals may compete with each other for:

- food

- water

- mates

- territory (space in which to live and breed).

remember >>
Plants make their own food by photosynthesis, so they do not compete for food.

exam tip >>
You may be given information about a particular habitat and be asked to suggest the factors that organisms are competing for there.

B Adaptation

1 Organisms have adaptations or particular features that allow them to survive in their environment.

2 **key fact** Predators may be warned or put off by adaptations such as warning colours, thorns or poisons. For example, wasps have black and yellow warning stripes and a sting.

3 **key fact** Animals and plants are adapted in different ways for living in cold or hot environments.

adaptation	Arctic fox	desert rat
body size	large – gives a small surface area compared to its volume, so heat is lost more slowly	small – gives a large surface area compared to its volume, so heat is lost more quickly
insulating layer of fat	thick	thin
insulating layer of fur	thick	thin
camouflage	white	brown

exam tip >>

Be prepared to suggest how a particular organism is adapted to its surroundings. You will probably be given information to help.

>> practice questions

1 State three things that animals may compete for.

2 Explain why grass under a leafy tree does not grow very well.

3 Suggest why the snowshoe hare has white fur in the winter and brown fur in the summer.

Reproduction

- Chromosomes, found in the nucleus of the cell, carry genes.
- Different genes control different characteristics of an organism.
- Sexual reproduction leads to variation in offspring.
- Asexual reproduction produces clones.

A Genes and chromosomes

1. DNA is deoxyribose nucleic acid. It is the chemical that carries genetic information. This information is passed from parents to their offspring in DNA.

2. A gene is a section of DNA that carries the code for a particular protein.

3. **key fact** Chromosomes are found in the nucleus of a cell. They contain many genes. Different genes control the development of different characteristics in an organism.

exam tip >>
You don't need to know the structure of DNA for your exams.

B Sexual reproduction

1. Sexual reproduction involves two parents.

2. Sex cells are called gametes. In humans:
   - eggs are the female gametes
   - sperm are the male gametes.

3. **key fact** In sexual reproduction, male and female gametes join to make a new cell. This develops into a new individual. The joining or fusion of gametes is called fertilisation.

exam tip >>
Gametes have half the normal number of chromosomes. For example, human body cells have 46 chromosomes but human gametes have 23. Fertilisation produces a new cell with the normal number, in this case 46.

④ **key fact** Sexual reproduction leads to variety in the offspring. Half the genetic information in the offspring comes from one parent, and half comes from the other parent. So the genetic information from two parents is mixed. The offspring look a bit like each parent, but are not identical to them.

C Asexual reproduction

① Asexual reproduction involves just one parent.

② **key fact** There is no fusion of gametes in asexual reproduction. This means that there is no mixing of genetic information and the offspring are identical to the parent.

③ **key fact** Genetically identical individuals are called clones. The parent and offspring are all clones.

>> practice questions

1 Where are chromosomes found in the cell?

2 Explain why the offspring from sexual reproduction are not genetically identical to their parents.

3 Why does asexual reproduction produce clones?

Cloning and genetic engineering

- Plants can be cloned quickly and cheaply by taking cuttings.
- There are several methods of modern cloning.
- Genetic engineering involves transferring genes from one organism to another.

A Cloning plants

1 key fact Plants can be cloned by taking cuttings. New plants can be produced quickly and cheaply this way.

A branch from the parent plant is cut off. Its lower leaves are removed and the stem is planted in damp compost.

Plant hormones are used to encourage new roots to develop. The method is simple enough for most gardeners to do.

2 key fact Plants can also be cloned by tissue culture.

Small groups of cells are taken from a part of the plant. They are added to sterile agar jelly containing nutrients and plant hormones. These encourage the cells to divide. Once the cells have grown into small masses of tissue, other hormones are added to stimulate the growth of roots and stems. The plantlets are then transferred into trays where they develop into plants.

3 key fact Tissue culture is more expensive and more difficult to do than taking cuttings.

B Cloning animals

1 key fact Animals can be cloned by embryo transplants, fusion cell cloning or by adult cell cloning.

2 **key fact** With embryo transplants, a developing embryo is removed from a pregnant animal. This is at an early stage before the cells have become specialised. The cells are separated, grown in a laboratory and then transplanted into host mothers.

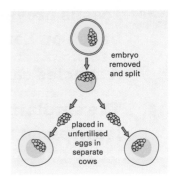
embryo removed and split

placed in unfertilised eggs in separate cows

3 The offspring are identical to each other. They are genetically related to the original pregnant animal. But they contain different genetic information from their host mothers.

4 Fusion cell cloning and adult cell cloning are similar to each other. The main stages are:

- the nucleus of an unfertilised egg cell is removed
- the nucleus from a different cell is injected into the egg cell
- the egg cell is implanted into the womb of a host mother where it develops.

5 **key fact** In fusion cell cloning the donor nucleus comes from an embryo. In adult cell cloning it comes from a body cell of an adult animal.

C Genetic engineering

1 Genetic engineering is also called genetic modification or GM. Genetically modified organisms are called GMOs.

2 **key fact** Genes are 'cut out' of the chromosome of one cell using enzymes and transferred to another cell. This is usually from one species to another.
For example:

- **bacteria have been produced that make human insulin**
- **crop plants have been developed that produce an insecticide.**

>> practice questions

1 Outline two ways in which plants can be cloned.

2 Explain why an animal produced by embryo transplant is not genetically identical to its host mother.

3 Geldings are male horses but they cannot reproduce. Describe how a valuable gelding racehorse could be cloned.

Extinction and evolution

- Fossils provide evidence that organisms have changed since life began on Earth.

- A species can become extinct for several reasons.

- The similarities and differences between species are analysed to learn about their evolutionary and ecological relationships.

A Fossils

1. A fossil is the remains of a prehistoric animal or plant, preserved in the Earth's crust. Usually, bones rather than soft parts are fossilised, but fossils can include footprints and other traces.

2. **key fact** Fossils give us evidence that living things have changed since life began on Earth. They may show that certain organisms have changed a lot, while others have changed very little.

3. Many fossil remains are of species that no longer exist. Such species have become extinct.

remember >>

Fossil footprints may show whether an animal walked on two legs or four, and how fast it was walking.

B Extinction

1. The fossil record shows that many species have existed on Earth in the past but no longer exist. They have become extinct. Extinction happens when there are no living members of a certain species left.

2. **key fact** Extinction may be caused by factors such as:

- **changes to the environment**
- **new competitors**
- **new diseases**
- **new predators.**

exam tip >>

Notice the use of the word 'new'. Remember that competitors, disease or predators are likely to be present all the time. It is the idea of a change that is important.

C Evolution

① key fact The theory of evolution states that:

- simple life forms developed on Earth more than three billion years ago

- all species of living things have evolved from these life forms.

② The fossil record provides evidence of evolution. But it has gaps. Not all organisms fossilise well. Many fossils will have been destroyed by natural movements of the Earth, or may not have been discovered yet.

③ There is a fairly complete evolutionary record of the horse. All the main stages of its evolution have been preserved as fossils. The horse evolved – over 60 million years – from a dog-sized creature, that lived in rainforests, into a tall animal adapted to living on the plains.

exam tip >>

You may be given pictures or other information about fossils, and asked to explain how it relates to the theory of evolution.

>> practice questions

1 What is a fossil?

2 A species of bird lives on an island. It nests on the ground. Explain why it might become extinct if rats escaped onto the island from a ship.

3 Outline the theory of evolution.

Natural selection

- Darwin's theory of natural selection explained how evolution happens.

- There have been other theories of evolution.

- A species can change rapidly because of mutation.

A Natural selection

1 Charles Darwin was an English naturalist. He studied variation in plants and animals during a five-year voyage around the world in the 19th century.
In 1859, Darwin published his ideas about evolution in a book called *On the Origin of Species*. He stated that evolution happens by natural selection.

2 **key fact** The main points about natural selection are:

- individuals in a species show a wide range of variation (they are not all the same)

- this variation is caused by differences in their genes

- individuals with characteristics most suited to the environment are more likely to survive and reproduce

- the genes that allow these individuals to be successful are passed to their offspring.

remember >>

Scientists did not know about genes in Darwin's time. But they did know that offspring inherited their characteristics from their parents.

B Criticism of natural selection

1 **key fact** Darwin's theory of natural selection was only gradually accepted. There are many reasons for this.

2 We may never be certain about how life began on Earth. No-one was there to observe it or record observations.

3 Darwin's ideas caused a lot of controversy at the time, which still continues. His ideas can be seen as conflicting with religious views about the creation of the world and the living things in it.

4 Living things alive today are very complex. This is true even of microorganisms. Some people find it difficult to accept that this could have happened because of natural selection.

C Lamarck's theory of evolution

(1) **key fact** There have been other theories of evolution.

(2) Jean-Baptiste Lamarck was a French scientist who developed an alternative theory in the 19th century. It involved two ideas:

- a characteristic which is used more becomes bigger and stronger. One that is not used eventually disappears
- any feature that is improved through use is passed to the offspring.

The table below explains the evolution of the giraffe's long neck using Lamarck's and Darwin's theories.

Lamarck	Darwin
A giraffe stretches its neck to reach food high up	A giraffe with a longer neck can reach food high up
The giraffe's neck gets longer because it is used a lot	The giraffe is more likely to get enough food to survive to reproduce
The giraffe's offspring inherit its long neck	The giraffe's offspring inherit its long neck

(3) Lamarck's theory would predict that all species should become complex over time, and simple organisms would become extinct. Lamarck's theory cannot explain why simple organisms still exist today, but Darwin's theory can.

>> practice questions

1 Outline the main points of natural selection.

2 Suggest a reason why Darwin's theory was accepted only gradually.

3 How can Darwin's theory of evolution explain why simple organisms such as bacteria exist, and have not all become extinct?

Pollution

- Human activities are using up raw materials and producing more waste.

- Waste can pollute the environment if it is not properly handled.

- Living organisms can be used as indicators of pollution.

A Impact of humans

1 **key fact** The human population is increasing rapidly and the standard of living for many people is improving. This increases demands on the environment, including increases in:

- the use of non-renewable energy resources

- the use of raw materials, such as metal ores

- the amount of waste produced.

2 Humans are very successful competitors. We compete with other species for natural resources. Farming, quarrying and building reduce the amount of land available for plants and animals.

B Pollution

1 Human activities produce waste products, including solids, liquids and gases. These may be toxic and they may be difficult to dispose of. If waste is not handled properly it can cause pollution.

2 **key fact** The air can be polluted with gases such as sulfur dioxide, and smoke. Sulfur dioxide contributes to acid rain. This damages buildings, living things in rivers and lakes, and trees. Smoke blackens buildings and can make it difficult for us to breathe.

3 **key fact** The land can be polluted with toxic chemicals. These include herbicides (weedkillers) and pesticides used by farmers to improve their crop yields. The rain can wash these chemicals into rivers, lakes and seas.

(4) **key fact** Rivers, lakes and seas can be polluted with toxic chemicals from factories, fertilisers and pesticides from fields, and sewage.

exam tip >>

Look out for questions that might involve knowledge and understanding from another area of science. For example, most plastics are not biodegradable. They are often dumped in landfill sites.

C Indicators of pollution

(1) Living things are often very sensitive to pollution in their environment. It may be difficult for them to survive if certain polluting chemicals are there. Scientists can find out how polluted the environment is by analysing the species present.

(2) **key fact** Invertebrate animals such as insect larvae can be used to indicate pollution in water. Clean rivers and lakes contain more species, and bigger populations, of invertebrates than polluted rivers and lakes.

(3) **key fact** Lichens are a type of plant that grows on walls and trees. They absorb water and minerals over their entire surface area from the rain and the air. This makes them sensitive to air pollution. Lichens grow well in unpolluted air, but there may be very few growing in towns.

>> practice questions

1 Explain why the use of non-renewable energy resources is increasing.

2 Outline one way in which farming can pollute the environment.

3 Explain why walls in towns may have few lichens growing on them, but walls in the countryside may have many lichens.

4 Suggest why it is important for future generations that we recycle metals.

Global warming

- Greenhouse gases such as carbon dioxide and methane absorb heat energy in the atmosphere.

- Human activities are increasing the amounts of greenhouse gases in the atmosphere.

- Increasing amounts of greenhouse gases may be causing global warming.

- Global warming may cause climate change and a rise in sea level.

A Greenhouse effect

1 The Earth's surface radiates thermal energy. Without an atmosphere this thermal energy would escape into space and the planet would be much colder than it is.

2 **key fact** Carbon dioxide and methane are greenhouse gases. They absorb most of the thermal energy radiated by the Earth.

3 **key fact** Some of this energy is radiated back to the Earth. It keeps the planet warmer than it would be without the greenhouse gases. This is called the greenhouse effect.

B Global warming

1 Human activities are releasing large, extra amounts of greenhouse gases into the atmosphere.

2 **key fact** Increasing levels of greenhouse gases increase the greenhouse effect. This in turn increases the average temperature of the planet. This is called global warming.

3 **key fact** Global warming by only a few degrees Celsius may cause:

- a rise in sea levels
- large-scale changes to the weather, called climate change.

C Methane

① **key fact** Methane is released into the atmosphere because of farming. It is produced by rice paddy fields and by cattle.

② **key fact** The world's population is increasing. This means that there are increasing numbers of cattle and rice paddy fields. In turn, this means that increasing amounts of methane are released into the atmosphere.

D Carbon dioxide

① Trees are being cut down in tropical areas of the world. This deforestation provides timber for building, and land for farming.

② Deforestation leads to a loss of biodiversity. Some of the species that become extinct might have been useful to us in the future. For example, they might have been a source of new medicines.

③ Deforestation also adds to the amount of carbon dioxide in the atmosphere, leading to global warming:

- there are fewer trees left to absorb carbon dioxide from the atmosphere
- fallen trees may be burnt, and this releases carbon dioxide
- decay microorganisms release carbon dioxide.

>> practice questions

1 What is a greenhouse gas?

2 Describe one problem that could be caused by global warming.

3 Explain how farming can increase the amount of methane and carbon dioxide released into the atmosphere.

Atoms and elements

- All substances are made from tiny particles called atoms.

- An element is a substance made from just one sort of atom.

- The elements are shown in the periodic table.

- Groups in the periodic table contain elements with similar properties.

A Atoms

1 All substances are made of atoms.

2 **key fact** There is a small nucleus at the centre of each atom, with even smaller particles called electrons arranged around it.

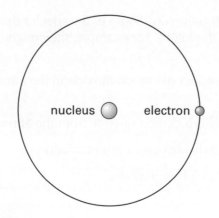

nucleus ○   electron ○

B Elements

1 **key fact** There are about 100 different elements.
Each element is made of just one sort of atom.
Different elements are made from different sorts of atoms.

2 Each element is given a chemical symbol. For example, the symbol for magnesium is Mg and the symbol for oxygen is O.

> **exam tip >>**
>
> Chemical symbols for the elements are made from one, two or three letters. The first letter is always a capital letter, and the next one or two are lower case.

C The periodic table

1 The periodic table is a chart showing information about the elements.

2 **key fact** The vertical columns in the periodic table are called groups. The elements in a group have similar properties to each other.

remember >>

You won't find substances like water and carbon dioxide in the periodic table because they are compounds, not elements.

Group 1 is called the alkali metals, group 2 is called the alkaline earth metals, group 7 is called the halogens, and group 0 is called the noble gases.

This line divides the metals from the non-metals.

transition metals

atomic number

H																		He
Hydrogen 1																		Helium 2

1	2											3	4	5	6	7	0
Li Lithium 3	Be Beryllium 4											B Boron 5	C Carbon 6	N Nitrogen 7	O Oxygen 8	F Fluorine 9	Ne Neon 10
Na Sodium 11	Mg Magnesium 12											Al Aluminium 13	Si Silicon 14	P Phosphorus 15	S Sulfur 16	Cl Chlorine 17	Ar Argon 18
K Potassium 19	Ca Calcium 20	Sc Scandium 21	Ti Titanium 22	V Vanadium 23	Cr Chromium 24	Mn Manganese 25	Fe Iron 26	Co Cobalt 27	Ni Nickel 28	Cu Copper 29	Zn Zinc 30	Ga Gallium 31	Ge Germanium 32	As Arsenic 33	Se Selenium 34	Br Bromine 35	Kr Krypton 36
Rb Rubidium 37	Sr Strontium 38	Y Yttrium 39	Zr Zirconium 40	Nb Niobium 41	Mo Molybdenum 42	Tc Technetium 43	Ru Ruthenium 44	Rh Rhodium 45	Pd Palladium 46	Ag Silver 47	Cd Cadmium 48	In Indium 49	Sn Tin 50	Sb Antimony 51	Te Tellurium 52	I Iodine 53	Xe Xenon 54
Cs Caesium 55	Ba Barium 56	La Lanthanum 57	Hf Hafnium 72	Ta Tantalum 73	W Tungsten 74	Re Rhenium 75	Os Osmium 76	Ir Iridium 77	Pt Platinum 78	Au Gold 79	Hg Mercury 80	Tl Thallium 81	Pb Lead 82	Bi Bismuth 83	Po Polonium 84	At Astatine 85	Rn Radon 86
Fr Francium 87	Ra Radium 88	Ac Actinium 89	Rf Rutherfordium 104	Db Dubnium 105	Sg Seaborgium 106	Bh Bohrium 107	Hs Hassium 108	Mt Meitnerium 109	Ds Darmstadtium 110	Rg Roentgenium 111							

elements 112 to 116 have also been reported

>> practice questions

1 Describe the structure of the atom.

2 What is an element?

3 Explain what a group is in the periodic table.

Reactions and compounds

- Atoms join together to form compounds in chemical reactions.

- Chemical bonds join atoms together.

- A chemical formula shows the number of atoms of each element in a compound.

- A balanced chemical equation shows what happens to each substance in a chemical reaction.

A   Chemical reactions

1 In a chemical reaction, the substances that react together are called reactants and the substances that are produced are called products.

2 **key fact** In a chemical reaction, the atoms in the reactants join together in a different way than before to make the products.

3 The joins between atoms are called chemical bonds. They form when electrons move from one atom to another, or are shared between two atoms.

exam tip   >>

You don't need to know any more than this about chemical bonds for GCSE Science. You will find out more about them if you study GCSE Additional Science.

B   Chemical formulae

>> **key fact** A chemical formula shows the number of each type of atom in a compound.

For example:

- H_2O is the formula for water. It shows that a molecule of water contains two hydrogen atoms and one oxygen atom.

- CO_2 is the formula for carbon dioxide. It shows that a molecule of carbon dioxide contains one carbon atom and two oxygen atoms.

remember >>

The numbers in chemical formulae are subscripts. So CO_2 is correct but CO^2 is not.

C Conservation of mass

>> **key fact** No atoms are lost or made during a chemical reaction.

This means that:

- the mass of the products is the same as the mass of reactants
- balanced chemical equations show all the atoms involved.

No atoms are lost or made when carbon reacts with oxygen to make carbon dioxide. They just join together in a different way than before.

D Balanced equations

① **key fact** In a balanced equation, the reactants are written on the left of the arrow and the products on the right of it.

② The equation is balanced by adjusting the same number of molecules on one or both sides of the arrow. This is done so that both sides have the same number of atoms of each element. For example:

- $C + O_2 \rightarrow CO_2$ is balanced because there is one C and two Os on both sides. The equation shows that one atom of carbon reacts with one molecule of oxygen to make one molecule of carbon dioxide.

- $Cu + O_2 \rightarrow CuO$ is not balanced. It is balanced by writing a number 2 against Cu and against CuO:

  $2Cu + O_2 \rightarrow 2CuO$ is balanced because there are now two Cus and two Os on both sides.

>> practice questions

1 6 g of carbon reacts with 16 g of oxygen. What mass of carbon dioxide is formed?

2 The chemical formula of glucose is $C_6H_{12}O_6$. What information does it give you?

3 Explain what happens to electrons so that chemical bonds form.

4 Balance this equation: $C + CuO \rightarrow CO_2 + Cu$.

Limestone

- Limestone contains calcium carbonate. It is used as a building material and as a raw material for making other substances.

- Calcium carbonate decomposes when heated, forming calcium oxide and carbon dioxide.

- Calcium oxide reacts with water to form calcium hydroxide.

A Limestone

1 **key fact** Limestone is a common rock found in many parts of the UK. It is mostly calcium carbonate, $CaCO_3$.

2 Limestone is removed from the ground in quarries.

3 Blocks of limestone are used as a building material. Crushed limestone chippings are used for road building.

B Thermal decomposition

1 **key fact** Calcium carbonate breaks down to form calcium oxide and carbon dioxide when it is heated strongly:

calcium carbonate → calcium oxide + carbon dioxide

$$CaCO_3 \rightarrow CaO + CO_2$$

This reaction is called thermal decomposition.

2 Other metal carbonates react in a similar way. For example, green copper carbonate decomposes when heated to form black copper oxide:

copper carbonate → copper oxide + carbon dioxide

$$CuCO_3 \rightarrow CuO + CO_2$$

C Quicklime and slaked lime

1 Calcium oxide is also called quicklime. Its chemical formula is CaO.

2 Calcium oxide reacts with water to produce calcium hydroxide, also called slaked lime. The formula of calcium hydroxide is $Ca(OH)_2$.

3 **key fact** Using common names instead of chemical names, this is how limestone is changed by chemical reactions:

- limestone $\xrightarrow{\text{heat}}$ quicklime + carbon dioxide

- quicklime + water $\rightarrow$ slaked lime.

D Uses of limestone and its products

1 Limestone is:

- used to remove impurities when iron is extracted from iron ore

- heated with sand and sodium carbonate to make glass

- heated with clay to make cement.

2 Cement is:

- mixed with sand and water to make mortar

- mixed with sand, water and crushed rock to make concrete.

3 Limestone, quicklime and slaked lime are used to reduce the acidity of fields and lakes affected by acid rain.

>> practice questions

1 Limestone quarries are often located in countryside areas, which may also be tourist areas. Outline some of the advantages and disadvantages of quarrying limestone.

2 Suggest why the demand for limestone increases when a large building project starts, such as the 2012 London Olympic Games.

Ores and iron

- Ores are rocks that contain useful amounts of metal.

- Chemical reactions are needed to extract most metals from their ores.

- Metals that are less reactive than carbon, such as iron, are extracted from their ores by reduction with carbon.

A Ores

① **key fact** Ores are naturally occurring rocks that contain enough metal to make it economical to extract the metal.

② Common ores include haematite (iron ore) and bauxite (aluminium ore).

③ High-grade ores contain a lot of the desired metal, and low-grade ores contain very little of it. Many of the world's high-grade ores have already been mined.

④ There is a limited supply of ores. Recycling metals after use reduces the need to mine ores and it saves energy.

B Methods of extraction

① **key fact** Unreactive metals are found in the Earth's crust as the metal itself. Gold is like this. But most metals need chemical reactions to extract them from their ores.

② Many metals are found in their ores as oxides. The metals are extracted by removing the oxygen. This is called reduction.

③ The method of chemical extraction depends on how reactive the metal is:

- metals that are less reactive than carbon can be extracted from their oxides by reduction with carbon or carbon monoxide

- metals that are more reactive than carbon are extracted using electricity.

	metal		method of extraction
most reactive	potassium sodium calcium magnesium aluminium	most difficult to extract	extracted using electricity (electrolysis)
	carbon		
	zinc iron tin lead copper		extracted using carbon
	silver		
least reactive	gold platinum	easiest to extract	found in native state

Part of the reactivity series of metals and the method of extraction.

C Iron

1 **key fact** Iron is less reactive than carbon, so it can be extracted from iron oxide by reduction using carbon.

2 Iron is extracted from iron oxide in a large container called a blast furnace. These are the equations for one of the reactions that take place there:

iron oxide + carbon → iron + carbon dioxide

$2Fe_2O_3 + 3C \rightarrow 4Fe + 3CO_2$

3 **key fact** Iron straight from the blast furnace contains about 96% iron. The remaining 4% is impurities such as carbon. These make the iron hard but brittle. This means it breaks easily when stretched or bent, so it has few uses. It may be used for some cooking stoves.

exam tip >>

You don't need to know any details of the blast furnace for your exams.

>> practice questions

1 What is an ore?

2 Explain why you could extract lead from lead oxide by reduction with carbon, but you could not extract aluminium from aluminium oxide this way.

3 Why does iron straight from the blast furnace have few uses?

Alloys

- Alloys are mixtures of a metal with non-metals or other metals.

- Alloys can be designed to have properties that make them suited for certain uses.

- Smart alloys go back to their original shape after being bent.

A Pure metals

1 **key fact** The atoms in pure metals are regularly arranged. Layers of atoms can slide over each other.

2 **key fact** Iron straight from the blast furnace is brittle because of its impurities. If all the impurities were removed, the pure iron would be soft and easily shaped. This is because layers of iron atoms could slide over each other.

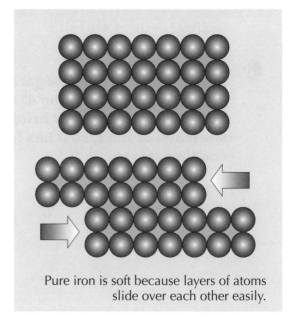

Pure iron is soft because layers of atoms slide over each other easily.

B Steel

1 **key fact** Pure iron would be too soft for many uses, so most iron is converted into steel.

2 Steel is made by removing most of the carbon from iron from the blast furnace, then adding other metals to make alloys.
There are different types of steel, including:

- low carbon steel that is easily shaped – useful for making car body panels

- high carbon steel that is hard – useful for making tools

- stainless steel that is resistant to corrosion – useful for sinks and cutlery.

C Other alloys

1 **key fact** Alloys contain different-sized atoms. These alter the regular arrangement of atoms so that the layers cannot slide over each other easily. This means that alloys are harder than the pure metal.

2 Different alloys can be designed with different properties. This makes them useful for particular purposes, and many everyday metals are alloys. For example:

- copper is alloyed with zinc to make brass, which is used in electrical fittings
- gold is alloyed with copper for use in jewellery
- aluminium is alloyed with copper and other metals to make duralumin, which is used in aircraft parts.

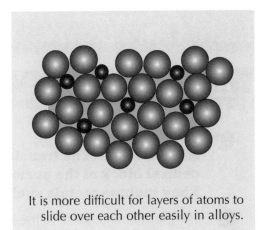

It is more difficult for layers of atoms to slide over each other easily in alloys.

D Smart alloys

1 **key fact** Smart alloys are mixtures of metals such as nickel and titanium that have unusual properties. For example, they spring back into shape after being bent.

2 Shape memory alloys are smart alloys which can be twisted and bent, but snap back into shape when they are warmed up.

>> practice questions

1 What is an alloy?

2 Explain why alloys are harder than the pure metals they contain.

3 Suggest why high carbon steel would not be suitable for making car body panels.

Copper, titanium and aluminium

- The elements in the middle block of the periodic table are called the transition metals.

- Many of the transition metals are useful as construction metals.

- Supplies of copper ore are limited.

- It is expensive to extract titanium and aluminium.

A The transition metals

1 key fact The transition metals are the elements in the central block of the periodic table, between group 2 and group 3. They include metals such as iron, copper and titanium.

2 The transition metals have the typical properties of metals. They:

- are good conductors of heat

- are good conductors of electricity

- can be hammered or bent into shape.

transition metals

3 Many transition metals are strong and tough. They are useful as structural materials, for example for making office blocks, bridges, vehicles and ships.

B Copper

1 key fact Most of the high-grade copper ores have already been mined, so they are in limited supply. Most copper is now extracted from low-grade ores. These contain relatively little copper. This means that a lot of waste is produced when they are mined and the copper is extracted.

2 Copper can be extracted by passing electricity through dilute solutions of copper compounds, such as copper sulfate. This is called electrolysis.

3 Electrolysis is expensive because large amounts of electricity are needed. But other metals are recovered in the waste sludge, such as silver. This helps to offset the cost, but not the environmental impact.

(4) New ways to extract copper are being researched, including using bacteria.

The table below shows some uses of copper and reasons for using it.

use of copper	reasons for using copper
plumbing (water pipes and fittings)	does not react with water easily bent into shape without breaking
electrical cables	easily made into wires a good conductor of electricity

C Titanium

(1) Titanium is a transition metal. Its surface is protected from corrosion by a layer of titanium oxide. It is strong and has a low density, so objects made from titanium are lightweight for their size. Titanium is used for aircraft parts and artificial hip joints.

(2) **key fact** Titanium cannot be extracted by reducing titanium oxide using carbon. It is expensive to extract titanium because several steps are needed, and each needs a lot of energy.

D Aluminium

(1) Aluminium is not a transition metal. It is found in group 3. But, like titanium, its surface is protected from corrosion by a layer of its oxide, and it has a low density.

(2) **key fact** Aluminium is more reactive than carbon, so it cannot be extracted by reducing aluminium oxide using carbon. It is expensive to extract aluminium because it is extracted by electrolysis, which needs a lot of energy.

remember >>

It is better to recycle metals than to extract them from their ores. It saves energy, preserves the limited supplies of ores, and has less impact on the environment.

>> practice questions

1 **What makes copper useful for plumbing?**

2 **Explain why it is expensive to extract titanium and aluminium from their ores.**

3 **Scientists have discovered a new way to extract titanium in one step. Suggest why this should make titanium more widely available in the future.**

Hydrocarbons

- Crude oil is a mixture of hydrocarbons.

- Hydrocarbons are compounds of hydrogen and carbon.

- Alkanes are hydrocarbons with the general formula C_nH_{2n+2}.

- The different hydrocarbons in crude oil can be separated by fractional distillation.

A Crude oil

1 Crude oil is a fossil fuel.

2 **key fact** Crude oil contains a large number of hydrocarbons. These are compounds containing hydrogen and carbon only.

B Alkanes

1 **key fact** Alkanes are the simplest hydrocarbons. The general formula of alkanes is C_nH_{2n+2}. This means that, for example, the molecular formula for ethane is C_2H_6.

2 Alkanes are saturated compounds. This means that all the bonds in alkanes are single bonds. The atoms in an alkane and how they are joined together can be shown by a structural formula.

name	molecular formula	structural formula
methane	CH_4	H | H—C—H | H
ethane	C_2H_6	H H | | H—C—C—H | | H H
propane	C_3H_8	H H H | | | H—C—C—C—H | | | H H H
butane	C_4H_{10}	H H H H | | | | H—C—C—C—C—H | | | | H H H H

The molecular formulae and structural formulae of the first four alkanes.

C Fractional distillation

1 The longer the alkane, the higher its boiling point. The different boiling points of alkanes allow crude oil to be separated into fractions by fractional distillation. They are called fractions because they are only a part of the original crude oil.

2 **key fact** Fractional distillation works by heating the crude oil to evaporate it, then letting it condense at several different temperatures:

- **gases leave the top of the fractionating column**

- **liquids leave at various levels in the middle**

- **solids stay at the bottom.**

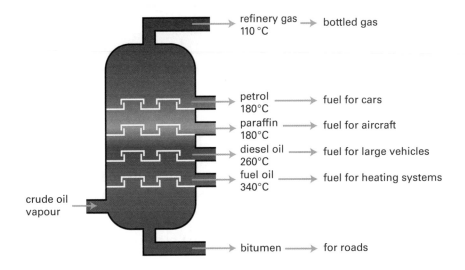

D Different fractions

>> key fact Shorter alkanes make better fuels than longer alkanes. This is because the shorter the hydrocarbon molecule:

- **the lower its boiling point (the more volatile it is)**

- **the runnier it is (its viscosity is lower)**

- **the easier it is to ignite.**

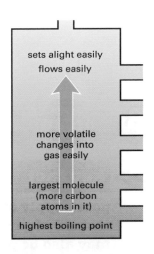

>> practice questions

1   What is a hydrocarbon?

2   Write the molecular formula of hexane, which contains six carbon atoms.

3   Explain how crude oil can be separated into fractions.

4   Explain why fractions from the top half of the fractionating column are more useful as fuels than those from the bottom half.

Burning fuels

- Hydrocarbons burn in air to produce water vapour and carbon dioxide.

- Many fuels contain some sulfur, which produces sulfur dioxide when the fuel is burned.

- The products from burning fuels damage the environment.

- Sulfur can be removed from fuels before their use, and sulfur dioxide can be removed from waste gases.

A Complete combustion

1. Fuels react with oxygen in the air when they burn. The reaction is called combustion. When there is plenty of air, burning is called complete combustion.

2. **key fact** The hydrogen in hydrocarbons forms water vapour when the fuel burns. The carbon in coal and hydrocarbons forms carbon dioxide when the fuel burns in plenty of air.

3. The general word equations for complete combustion are:

    - carbon + oxygen → carbon dioxide   (for coal)

    - hydrocarbon + oxygen → water vapour + carbon dioxide.

4. The balanced symbol equations for some hydrocarbons burning are:

    - $CH_4 + 2O_2 \rightarrow CO_2 + 2H_2O$ (for natural gas)

    - $C_4H_{10} + 6\frac{1}{2}O_2 \rightarrow 4CO_2 + 5H_2O$ (for butane).

B Incomplete combustion

>> **key fact** Incomplete combustion happens when there is a poor supply of air. Poisonous carbon monoxide gas is produced instead of carbon dioxide. Particles of carbon (smoke and soot) may also be produced.

C Pollution from burning fuels

1 **key fact** Carbon dioxide is a greenhouse gas. It prevents heat energy escaping from the atmosphere and contributes to global warming. If there is less demand for fossil fuels, the amount of carbon dioxide released will go down. This will reduce global warming.

2 **key fact** Particles cause global dimming. They increase the cloud cover, which reduces the amount of sunlight reaching the ground. Laws requiring smokeless fuels help to reduce the amount of particles released.

3 **key fact** Alternative fuels such as ethanol and hydrogen may reduce these pollutants. Ethanol produced by fermentation of sugars can be a carbon neutral fuel. Water vapour is the only product formed when hydrogen burns.

> **remember >>**
>
> Ethanol will not be a carbon neutral fuel if some of the energy needed to produce it comes from fossil fuels.

> **exam tip >>**
>
> Don't confuse greenhouse gases and global warming with damage to the ozone layer.

D Acid rain

1 **key fact** Sulfur dioxide from burning fuels contributes to acid rain. Fuels can be processed to remove the sulfur before they are used.

2 **key fact** Sulfur dioxide can be removed from waste gases, for example at power stations. One way to do this involves spraying powdered limestone into the waste gases. A reaction happens that produces gypsum, which is used to make plasterboard.

>> practice questions

1 Name five products of combustion.

2 What is the difference between complete combustion and incomplete combustion?

3 Some scientists believe that global dimming has helped to reduce global warming over the last few decades. Suggest what could happen as technology improves and fewer particles are emitted when fuels are used.

- Alkenes are hydrocarbons that contain a carbon–carbon double bond.

- Alkenes have the general formula C_nH_{2n}.

- Alkenes can be produced by cracking crude oil fractions.

- Ethanol can be produced by reacting ethene with steam.

A Alkenes

① key fact Alkenes are hydrocarbons with the general formula C_nH_{2n}. This means that, for example, the molecular formula for ethene is C_2H_4.

② key fact Alkenes are unsaturated compounds. This means that they contain a double bond.

③ Alkenes decolourise brown bromine water, but alkanes do not.

ethene C_2H_4 propene C_3H_6 butene C_4H_8

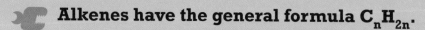

The molecular formulae and structural formulae of the first three alkenes.

exam tip >>

Take care that your letter *e* does not look a letter *a*, otherwise the word *alkene* will look like *alkane* when you write it.

B Cracking

① key fact The demand for shorter alkanes is greater than can be supplied by fractional distillation alone. Cracking is a process in which long alkanes are broken down into shorter alkanes and alkenes. The shorter alkanes are useful as fuels.

② key fact Cracking is an example of a thermal decomposition reaction. Alkanes are heated so that they vaporise, which is then passed over a hot catalyst, such as aluminium oxide.

3 Here is an example of the equations for a cracking reaction:

decane → octane + ethene

$C_{10}H_{22} \rightarrow C_8H_{18} + C_2H_4$

remember >>

Another thermal decomposition reaction is the breakdown of metal carbonates to metal oxides and carbon dioxide.

C Ethanol

1 **key fact** Ethene reacts with steam at high temperatures and pressures to form ethanol, C_2H_5OH. A catalyst of phosphoric acid is needed, too.

2 Ethanol is the only product in the reaction:

ethene + steam → ethanol

$C_2H_4 + H_2O \rightarrow C_2H_5OH$

Diagram to illustrate how ethanol is produced by reacting ethene with steam.

3 Ethanol can be produced from renewable resources by fermentation of sugars from plant materials. It is often called bioethanol.

4 **key fact** The use of bioethanol as a fuel reduces the demand for fossil fuels. But it uses farm land that could have been used to grow crops to feed people.

>> practice questions

1 **What is cracking and why is it carried out?**

2 **Write the molecular formula for hexene, an alkene with six carbon atoms.**

3 **Suggest why the alcohol for beer and wine is ethanol from fermentation, rather than ethanol from the reaction between ethene and steam.**

Polymers

- Alkenes can join together to make very long molecules called polymers.

- The properties of polymers depend on what they were made from and the conditions used.

- Polymers have very many uses but can be difficult to dispose of once they have been used.

A Monomers and polymers

1   Alkenes can join together because they contain a carbon–carbon double bond.

2   **key fact** When alkenes join together to make polymers, they are called monomers. Different monomers make different polymers. For example, ethene makes poly(ethene) and propene makes poly(propene).

3   Polymers have many uses, including:

- plastic bags

- dental fillings

- shape memory polymers for shrink-wrap packaging

- hydrogels for making soft disposable nappies and contact lenses.

$$
\begin{array}{ccccc}
CH_3 & H & CH_3 & H & CH_3 & H \\
| & | & | & | & | & | \\
C = C & + & C = C & + & C = C \\
| & | & | & | & | & | \\
H & H & H & H & H & H
\end{array}
$$

$$\downarrow$$

$$
\begin{array}{cccccc}
CH_3 & H & CH_3 & H & CH_3 & H \\
| & | & | & | & | & | \\
-C - C - & & C - C - & & C - C - \\
| & | & | & | & | & | \\
H & H & H & H & H & H
\end{array}
$$

Propene monomers polymerise to make poly(propene).

B Properties of polymers

1   **key fact** The properties of polymers depend on what they were made from and the conditions used.

2 Polymer molecules can have more or fewer branches, depending on the conditions used to make them. The properties of poly(ethene) depend upon the amount of branching.

polymer	branches	relative strength	maximum useable temperature /°C
LDPE low-density poly(ethene)	many	weak	85
HDPE high-density poly(ethene)	few	strong	120

3 Substances called plasticisers allow polymer molecules to slide over each other more easily. PVC with plasticisers is soft and flexible. It is used for floor coverings and raincoats. But unplasticised PVC or uPVC is hard. It is used to make pipes and window frames.

4 Poly(ethenol) is a polymer that dissolves in water to make slime. Its viscosity depends on the proportions of the ingredients used to make it.

exam tip >>

You may be given information in the exam about the properties of some polymers. Use this with your knowledge and understanding of science to answer the question.

C Disposing of polymers

1 **key fact** Biodegradable polymers can be broken down by microorganisms.

2 **key fact** Many polymers are not biodegradable. They take a very long time to break down. This makes them difficult to dispose of once they have been used.

>> practice questions

1 Name the polymer formed from chloroethene monomers.

2 Use information from the table to explain which polymer, LDPE or HDPE, would be best for making disposable hot drinks cups.

3 Suggest why the cost of polymers often depends on the cost of crude oil.

Plant oils

- Some fruit, seeds and nuts contain vegetable oils that can be extracted.
- Vegetable oils are an important part of our diet, and can be used as fuels.
- Emulsions are mixtures of oils and water, as used in salad dressings.
- Unsaturated vegetable oils can be hardened by reacting them with hydrogen.

A Extracting vegetable oils

1 **key fact** Vegetable oils are natural oils found in some fruit, seeds and nuts.

2 **key fact** Oils can be extracted by:
- crushing the plant material
- pressing the crushed material to squeeze out the oil
- removing impurities such as water.

3 **key fact** Oils can also be extracted by:
- dissolving the oil in a solvent
- distilling to remove the solvent.

B Uses of vegetable oils

1 **key fact** Vegetable oils are an important part of a balanced diet. They are a source of energy. They also contain vitamins and other nutrients.

2 **key fact** Vegetable oils are also used as fuels for vehicles. For example, biodiesel is extracted from oilseed rape plants or from used cooking oil.

> **remember >>**
>
> A carbon-neutral fuel does not add extra carbon dioxide to the atmosphere when it is used.

The table below shows some advantages and disadvantages of using vegetable oils as fuels.

advantages	disadvantages
reduces the use of fossil fuels	reduces the amount of land available for growing food crops
can be a carbon-neutral fuel	energy from fossil fuels may be needed to produce and transport it

C Vegetable oils in food

1 **key fact** Vegetable oils do not dissolve in water, but they can form mixtures called emulsions. These are more viscous (thicker and less runny) than oil or water.

2 Emulsifiers are substances that prevent the oil and water in an emulsion from separating again.

3 **key fact** Emulsions improve the appearance and texture of food. They are used in foods such as ice cream and salad dressings.

D Hardening vegetable oils

1 **key fact** Unsaturated vegetable oils contain carbon–carbon double bonds. These can be detected using bromine or iodine. The more double bonds, the more bromine or iodine reacts.

2 **key fact** Unsaturated vegetable oils are usually liquids at room temperature. They can be hardened by reacting them with hydrogen at about 60 °C. A nickel catalyst is needed too.

Hydrogen adds across a carbon–carbon double bond, forming a carbon–carbon single bond.

3 **key fact** Hardened oils are more saturated. They are called hydrogenated vegetable oils. They have higher melting points than unsaturated vegetable oils. This means that they:

- are solid at room temperature
- are useful in margarine
- can be used to make pastries and cakes.

>> practice questions

1 Outline the steps to extract a vegetable oil.

2 Why do we need vegetable oils in our diet?

3 What is an emulsion, and where might you find one in food?

4 Describe how vegetable oils are hardened, and explain why this is done.

Food additives

- Additives are put into food to improve its taste, appearance and shelf-life.

- Permitted additives are given E numbers.

- Additives can be identified by chemical analysis.

- Chromatography can be used to detect and identify artificial colours.

A Food additives

1 key fact Food additives are included in processed foods to improve:

- taste

- appearance

- shelf-life (how long the food lasts when stored).

2 key fact Food additives must be identified in the list of ingredients on the packet. Some are natural and some are artificial.

3 key fact Some additives have an E number. This means that they have been tested for safety, passed for use in food and licensed by the European Union.

The table below shows some food additives and why they are used in food.

type of additive	example	typical use
colouring	E102 (tartrazine)	orange colouring for soft drinks and sweets
preservative	E210 (benzoic acid)	stops harmful microorganisms growing
emulsifier	E322 (lecithin)	allows oil and water to form emulsions for ice cream and salad dressings
sweetener	E951 (aspartame)	low-calorie food and drink sweetening

B Risks of food additives

1 Some additives can cause allergic reactions in some people. For example, the UK Food Standards Agency has strict limits on the amount of colourings allowed in food. They are banned from baby foods.

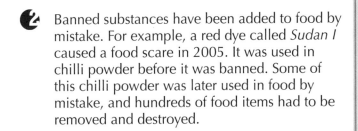

Banned substances have been added to food by mistake. For example, a red dye called *Sudan I* caused a food scare in 2005. It was used in chilli powder before it was banned. Some of this chilli powder was later used in food by mistake, and hundreds of food items had to be removed and destroyed.

exam tip   >>

You may be asked to evaluate the benefits and drawbacks of food additives using information about them.

C Analysing food

1 **key fact** Additives in food can be identified by chemical analysis.

2 **key fact** Chromatography is used to detect and identify artificial food colourings.

remember >>

Modern chromatography in analytical laboratories is automated to get fast, accurate results from tiny samples.

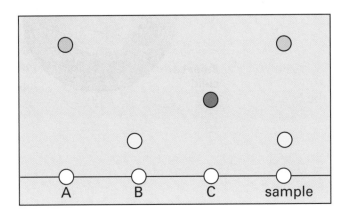

The sample in this chromatogram contains food colourings A and B, but not C.

>> practice questions

1 Give two reasons why processed food may contain additives.

2 How do you know if food contains additives?

3 Describe a risk of artificial food additives.

Structure of the Earth

- The Earth consists of a core, mantle and crust.

- The crust and upper part of the mantle are broken into tectonic plates.

- Tectonic plates move very slowly.

- Earthquakes and volcanoes happen where plates meet.

A Structure of the Earth

1 **key fact** The Earth consists of several layers, the core, mantle and crust.

2 The core consists of iron and nickel. The outer core is liquid but the inner core is solid.

3 The mantle is made from rock, but this can flow very slowly.

4 The crust is made from solid rock. It is the thinnest layer, about 50 km deep.

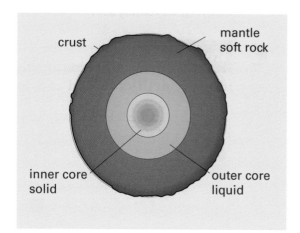

B Plate tectonics

1 **key fact** The Earth's crust and the upper part of the mantle are broken into huge pieces called tectonic plates.

2 **key fact** Convection currents in the mantle are driven by heat produced from radioactive processes deep inside the Earth. These currents cause the plates to move at a few centimetres per year.

3 **key fact** Where tectonic plates meet, the Earth's crust becomes unstable. The plates push against each other, or move over and under each other. Earthquakes and volcanic eruptions happen at the boundaries between plates.

4 Although areas of plate boundaries are known for earthquakes and volcanoes, it is difficult to predict exactly when these might happen and how bad they will be.

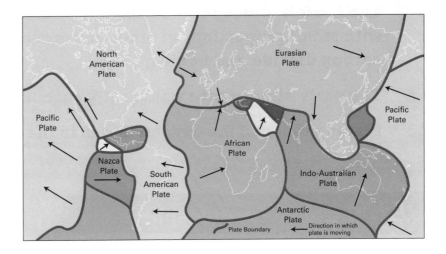

C Evidence for plate tectonics

1 **key fact** At one time scientists thought that mountains and other features on the Earth's surface happened because the crust shrank as the Earth cooled.

2 One problem with this theory is it does not explain why mountains only occur in certain places, and not all over the Earth's surface.

3 Alfred Wegener, in the early part of the last century, proposed the theory of continental drift involving tectonic plates. His evidence included:

- the east coast of South America and the west coast of Africa fit together like jigsaw pieces

- similar rocks and fossils are found on both continents.

This suggested that the two continents were once part of a single land mass which had moved apart.

4 **key fact** Wegener's theory explained why mountains occur in certain places, at plate boundaries. But he could not explain how tectonic plates could move. His theory was not generally accepted by scientists for many years, until sufficient evidence was collected.

>> practice questions

1 Describe the structure of the Earth.

2 State one problem with the idea that shrinkage of the crust caused mountains.

3 Outline two pieces of evidence for plate tectonics.

4 Where do earthquakes and volcanoes happen?

The atmosphere

The Earth's atmosphere is about 80% nitrogen and 20% oxygen, with small proportions of other gases such as water vapour, carbon dioxide and noble gases.

The early atmosphere was mostly carbon dioxide.

Plants produced the oxygen that is in the atmosphere.

A The modern atmosphere

>> **key fact** The Earth's atmosphere has stayed roughly the same for the last 200 million years:

- 78% nitrogen
- 0.9% argon (a noble gas)
- 21% oxygen
- 0.1% of other gases, such as carbon dioxide and water vapour.

exam tip >>

You may be asked to complete a pie chart to show the proportions of the main gases. Notice that nitrogen is about 80% and oxygen about 20%.

B The early atmosphere

1 The atmospheres of Mars and Venus today are mostly carbon dioxide with little or no oxygen. Theories about the history of the Earth suggest that the Earth's atmosphere was like this to begin with.

2 **key fact** During the first billion years after the Earth formed, intense volcanic activity released gases. These became the early atmosphere. This was mostly carbon dioxide with some water vapour, methane and ammonia.

3 As the Earth cooled, water vapour from the early atmosphere condensed to form the oceans.

remember >>

Theories about the evolution of the atmosphere must be able to explain why carbon dioxide decreased, and oxygen and nitrogen increased.

C Evolution of the atmosphere

1 key fact Plants evolved. Photosynthesis by plants produced the oxygen in the atmosphere and continues to do so today.

2 key fact Carbon dioxide decreased because it:

- dissolved in the oceans
- was taken in by plants and eventually became fossil fuels
- formed carbonates in sedimentary rocks such as limestone (calcium carbonate).

3 Methane and ammonia decreased because they reacted with oxygen. Nitrogen increased because of the reaction of ammonia with oxygen, and the action of certain bacteria.

D Carbon dioxide

>> key fact The level of carbon dioxide in the atmosphere is increasing. The burning of fossil fuels is adding carbon dioxide to the atmosphere faster than it can be removed.

>> practice questions

1 What was the main gas in the early atmosphere?

2 What is the main gas in the modern atmosphere?

3 Outline the reasons why carbon dioxide in the atmosphere decreased and oxygen increased.

4 Suggest why these ideas about the evolution of the atmosphere are described as theories, rather than facts.

Noble gases

- The noble gases are found in group 0 of the periodic table.

- They are all chemically unreactive.

- The noble gases are used in light bulbs and advertising signs.

- Helium is less dense than air, so it is used as a lifting gas in balloons.

A The noble gases

1 **key fact** The noble gases are the elements in group 0 of the periodic table.

2 Group 0 is found on the far right hand side of the periodic table. It includes:

- helium

- neon

- argon

- krypton

- xenon

- radon.

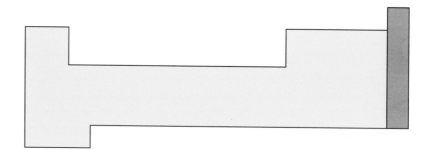

The noble gases are placed in the far right hand column of the periodic table.

B Properties of the noble gases

>> **key fact** The noble gases are all:

- **non-metals**

- **gases at room temperature**

- **chemically unreactive.**

exam tip >>

If an element's name ends in 'um' it is a metal. There is one exception: helium is a non-metal.

C Uses of the noble gases

① **key fact** Argon is used in filament lamps (ordinary light bulbs).

② There is a very thin metal wire inside filament lamps which gets hot and glows. When unreactive argon replaces air in the lamp, it stops the wire burning away.

③ **key fact** The noble gases glow when high voltage electricity is passed through them. They are used in electric discharge tubes, used for 'neon' advertising signs.

D Helium as a lifting gas

① **key fact** Helium is much less dense than air, so it can be used as a lifting gas. This means that balloons filled with helium float upwards.

② Helium is used in:

- party balloons
- weather balloons
- airships.

> **exam tip >>**
>
> Hydrogen is also much less dense than air. It was used in the past to fill airships, but hydrogen is flammable. The advantage of helium is that it is not flammable, reducing the risk of explosion.

>> practice questions

1 Name three noble gases.

2 Why is helium used as a lifting gas?

3 Why is argon used in filament lamps?

4 Explain why there are no naturally occurring compounds of the noble gases.

Heat transfer

- Heat can be transferred by radiation, conduction and convection.

- Conduction and convection involve particles.

- All objects emit and absorb infra red radiation, but different surfaces are better at doing this than others.

A Thermal energy

1 Heat energy is also called thermal energy.

2 The atoms in solids, liquids and gases are always vibrating and moving. Their movements increase when they absorb thermal energy.

3 **key fact** Thermal energy can be transferred from place to place by conduction, convection and radiation.

B Conduction

1 **key fact** Conduction of thermal energy needs particles.

2 **key fact** Conduction works best in solids because their particles are close together. It cannot happen through empty space.

3 Thermal energy is transferred from the hotter part of an object to the colder part. The particles in the hot part are vibrating more. These vibrations are passed on to the cooler particles next to them. In this way, energy spreads through the substance.

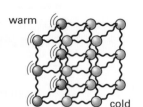
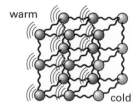

4 Metals are good heat conductors. Most non-metals, liquids and gases are poor heat conductors.

5 Insulators are poor conductors of heat. They keep things cold as well as hot. Air is a good insulator. Several layers of clothing will keep you warmer than one thick layer because the layers trap air between them.

C Convection

1 **key fact** Convection of thermal energy needs particles.

2 **key fact** Convection works in liquids and gases because their particles can move from place to place. It cannot happen in solids or empty space.

3
- When a liquid or gas warms up, it expands and becomes less dense.
- The warmer liquid or gas moves up through the cooler liquid or gas.
- It cools, contracts and becomes denser again, so it sinks.
- This creates a flow called a convection current.

D Radiation

1 **key fact** All objects emit (give off) and absorb (take in) thermal radiation. The hotter an object is, the more energy it emits.

2 **key fact** Thermal energy can be transferred as infra red radiation, also called thermal radiation. Transfer of thermal energy by radiation does not need particles. It can happen through empty space.

3 **key fact** Dark, dull (matt) surfaces are good emitters and absorbers of infra red radiation. Light, shiny surfaces are poor emitters and absorbers of infra red radiation.

E Factors affecting heat transfer

>> **key fact** The rate of transfer of thermal energy depends on:
- the type of material
- the shape of the object
- the size of the object
- the difference in temperature between the object and its surroundings – the bigger the difference, the faster the rate of transfer.

>> practice questions

1 Why can you feel thermal energy from the Sun, even though it is 150 million km away in space?

2 Explain why putting a lid on a cup of coffee keeps the coffee warmer for longer.

3 Suggest why satellites and spacecraft are often covered by sheets of shiny metal foil.

Efficiency

- Energy is measured in joules, J.

- The efficiency of a device is the proportion of the energy usefully transformed by the device.

A Energy transfers and transformations

1 key fact Energy is measured in joules, J.

2 key fact Energy cannot be created or destroyed.

3 key fact Energy can be transferred from place to place, or transformed from one type to another.

For example:

- thermal energy is transferred from one place to another by infra red radiation

- electrical energy is transferred from the power station to our homes through cables

- a battery transforms chemical energy into electrical energy

- a filament lamp transforms electrical energy into light energy.

B Useful transfers and transformations

1 key fact In any process or device, only some of the energy is usefully transferred or transformed. For example, a filament lamp transforms electrical energy into light energy, but it also transforms electrical energy into thermal energy.

remember >>
Wasted energy is not destroyed.

2 key fact The energy that is not usefully transferred or transformed is often called wasted energy.

3 key fact Energy that is transferred or transformed is eventually transferred to the surroundings as thermal energy. The surroundings warm up as a result. But the energy spreads out so much that it becomes very difficult to do anything useful with it.

For example: you should be able to identify the main energy wastages in different devices.

device	useful energy transformation	main energy wastage
filament lamp	electrical → light	electrical → thermal
car engine	chemical → kinetic	chemical → thermal chemical → sound

C Efficiency

① key fact **The efficiency of a device tells you what percentage of the energy used by the device is usefully transferred or transformed. It is calculated using this equation:**

$$\text{efficiency} = \frac{\text{useful energy transferred by device}}{\text{total energy supplied to device}}$$

② An ordinary filament lamp may transform 100 J of electrical energy to 10 J of light energy. The rest is transformed to thermal energy. Its efficiency is:

$$\text{efficiency} = \frac{\text{useful energy transferred by device}}{\text{total energy supplied to device}} = \frac{10}{100} = 0.1$$

An 'energy efficient' lamp may transform 100 J of electrical energy to 75 J of light energy. Its efficiency is 0.75 – it wastes much less energy as thermal energy.

exam tip >>

If you calculate efficiency and it is 1.0 or more, you have gone wrong. Efficiency is always less than 1.0 because some energy is wasted.

>> practice questions

1 What is 'wasted' energy?

2 What happens when energy is eventually transferred to the surroundings?

3 The efficiency of a coal-fired power station is 0.35 – how much electrical energy is produced from coal containing 100 MJ of chemical energy?

4 A television set transfers 200 J of electrical energy as 140 J of sound and light energy. What is its efficiency?

Cost of electricity

- The power of a device is the rate at which it transforms energy.

- Power is measured in watts, W, or kilowatts, kW.

- The total cost of the electricity used by a device depends upon its power, how long it is run for, and the cost of a unit of electricity.

A Energy and power

1 **key fact** The power of a device is the rate at which it transforms energy. The greater the power of a device, the more energy it transfers every second.

2 **key fact** Power is measured in watts, W. It can also be measured in kilowatts, kW.

3 A 2 kW electric kettle transforms energy at a greater rate than a 100 W electric lamp.

B Electrical energy transferred from the mains

1 **key fact** The electrical energy transferred from the mains supply depends on:

- the power of the device, and

- how long it is switched on.

2 **key fact** Energy is normally measured in joules, J. For mains electricity, electrical energy is measured in kilowatt-hours, kWh.

3 **key fact** The electrical energy transferred is calculated using this equation:

energy transferred	=	power	×	time
(kilowatt-hours, kWh)		(kilowatt, kW)		(hour, h)

④ *Example* A 2 kW electric kettle is switched on for 3 minutes. How much energy is transferred?

Answer Convert minutes to hours by dividing by 60:
3 minutes = 3 ÷ 60 = 0.05 h
energy transferred = power × time = 2 × 0.05 = 0.1 kWh

⑤ *Example* A 100 W electric lamp is switched on for 2 hours. How much energy is transferred?

Answer Convert W to kW by dividing by 1000:
100 W = 100 ÷ 1000 = 0.1 kW
energy transferred = power × time = 0.1 × 2 = 0.2 kWh

Notice that the lamp used twice as much electrical energy as the kettle in these examples.

remember >>

1 hour = 60 minutes. To convert from minutes to hours, divide by 60.

C Cost of electrical energy from the mains

① **key fact** The cost of electrical energy transferred from the mains supply depends on:

- **the energy used by the device, and**
- **the cost of a unit of electrical energy.**

② 1 kWh is 1 unit of electrical energy.

③ **key fact** The cost of electrical energy transferred is calculated using this equation:

total cost = number of kilowatt-hours × cost per kilowatt-hour

④ *Example* An electric fire uses 2 kWh of electrical energy. If 1 kWh costs 10p, how much does the fire cost to run per hour?

Answer total cost = number of kilowatt-hours × cost per kilowatt-hour
total cost = 2 × 10 = 20p

>> practice questions

1 What is the unit of power?

2 Write the equation that links electrical energy transferred from the mains with power and time.

3 How much electrical energy is transferred when a 250 W television set is used for 8 hours?

4 A 100 W electric lamp is switched on for 6 hours. If 1 kWh costs 10p, how much did the lamp cost to run?

The National Grid

- The National Grid transfers electricity from power stations to consumers.

- Step-up transformers increase the voltage, and step-down transformers reduce it.

- Transformers are used in the National Grid to reduce energy losses in the cables.

A The National Grid

1 **key fact** The National Grid transfers electricity from power stations to consumers.

2 The National Grid consists of many electrical cables, some underground and some overground, suspended from pylons. It is designed so that it can still supply electricity when demand increases suddenly, or a power station fails.

3 Power stations produce electricity at 25 000 V. Electricity is sent through the National Grid cables at 400 000 V, 275 000 V and 132 000 V. These high voltages are needed to reduce energy losses during transmission.

remember >>

You don't have to remember these voltages, but you do need to know why high voltages are used.

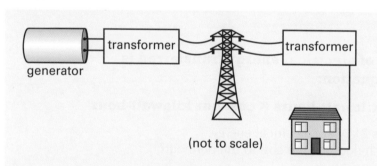

(not to scale)

There are many ways to generate electricity, including nuclear power and wind power.

B Energy losses

1 For a given rate of electrical energy transfer, the higher the voltage the lower the current.

2 Cables heat up when electricity is passed through them. When thermal energy is transferred from cables, energy is wasted. These losses must be reduced.

3 **key fact** The National Grid uses high voltages so that the current is reduced. This reduces the energy losses from the cables.

C Transformers

1 A transformer is an electrical device that changes the voltage of an alternating current, such as the mains electricity supply.

2 There are two main types of transformer:

- step-up transformers increase the voltage
- step-down transformers decrease the voltage.

3 The high voltages used by the National Grid reduce energy losses, but they are too dangerous to use in homes.

4 **key fact** **Step-up transformers increase the voltage from power stations so that energy losses from the cables are reduced.**

5 **key fact** **Step-down transformers reduce the voltage from the cables to safer levels for supply to homes.**

remember >>

The voltage of household electricity is about 230 V, but this is still dangerous.

>> practice questions

1 What is the National Grid?

2 Explain why electricity is transferred from power stations to consumers at high voltages.

3 Describe how transformers are used in the National Grid.

Generating electricity

- Electricity is generated in power stations.

- Most electricity is generated from fossil fuels or nuclear fuels.

- An energy source is used to produce steam from water, which is used to drive a turbine coupled to a generator.

A Energy sources

1. Most power stations use fossil fuels or nuclear fuels.

2. **key fact** Thermal energy is released from the fuel and used to heat water.

3. **key fact** Steam from the boiling water expands and pushes against the blades of a turbine.

4. **key fact** The turbine is connected to a generator. When the turbine spins, it turns the generator and electricity is produced.

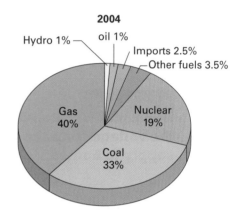

2004

Hydro 1% — oil 1%
Imports 2.5%
Other fuels 3.5%
Gas 40%
Nuclear 19%
Coal 33%

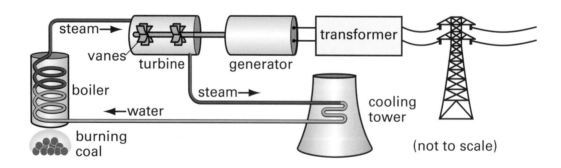

steam →
vanes
turbine
generator
transformer
boiler
steam →
← water
cooling tower
burning coal
(not to scale)

B Energy transfers

1 There are several energy transfers involved in generating electricity from fossil fuels or nuclear fuels. Energy is wasted at each step. The efficiency of a coal-fired power station is about 37%, and the efficiency of a nuclear power station is about 42%.

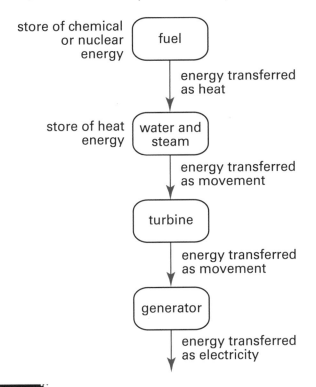

store of chemical or nuclear energy — fuel

energy transferred as heat

store of heat energy — water and steam

energy transferred as movement

turbine

energy transferred as movement

generator

energy transferred as electricity

2 **key fact** Power stations that use fossil fuels burn coal, natural gas or oil to produce thermal energy.

3 **key fact** Nuclear power stations use uranium or plutonium. These radioactive metals produce thermal energy because of nuclear reactions in them. The type of nuclear reaction is called nuclear fission.

exam tip >>

You don't need to know any details of nuclear reactions in GCSE Science. You will find out about them if you study GCSE Additional Science.

>> practice questions

1 Name three fossil fuels and two nuclear fuels.

2 Explain how thermal energy is used to produce electricity in a power station.

3 Suggest why the efficiency of power stations is quite low.

Renewable energy sources

> ◀⚡ Renewable energy sources include wind power, wave power, tidal power, hydroelectric power, solar power and geothermal energy.
>
> ◀⚡ Electricity can be generated from renewable energy sources.

A   Renewable energy sources

① **key fact** Renewable energy sources will not run out, as long as they are managed effectively.

② **key fact** Energy from wind, waves, tides and hydroelectric schemes can be used to drive turbines directly.

B   Wind power

① Wind turbines have huge blades mounted on a tall tower. The blades are connected to a generator. As the wind blows, it transfers some of its kinetic energy to the blades. These turn and drive the generator. Wind farms consist of several wind turbines grouped together in windy locations.

② **key fact** There are no fuel costs for a wind turbine and no polluting gases are produced.

③ **key fact** Wind farms are noisy and may spoil the view. If there is no wind, no electricity is generated.

C   Water power

① Waves at sea make the water level rise and fall. Wave machines use the kinetic energy in this movement to drive generators.

② The tides cause huge amounts of water to move in and out of a river estuary each day. A tidal barrage is a barrier built across a river estuary. It contains electricity generators, which are driven by water moving through pipes in the barrage as the tide changes.

③ Hydroelectric power schemes use the kinetic energy in moving water. A dam is built across a river valley. The water high up behind the dam contains gravitational potential energy. This is transferred to kinetic energy as the water moves down through pipes inside the dam. The moving water drives electricity generators.

④ **key fact** Tidal barrages and hydroelectric power stations are very reliable. They can be turned on quickly.

⑤ **key fact** Tidal barrages destroy the habitat of estuary species, such as wading birds. Dams flood farmland and people lose their homes. Rotting vegetation underwater releases methane, which is a greenhouse gas.

D Solar power

① **key fact** Solar cells convert light energy directly into electrical energy.

② **key fact** Solar cells are convenient for supplying electricity to small portable devices such as calculators, or to devices in remote areas such as road signs.

③ **key fact** Solar cells are expensive and inefficient, so the cost of their electricity is high. They do not work at night.

E Geothermal energy

① **key fact** Geothermal energy is thermal energy in hot rocks deep underground.

② **key fact** In some volcanic areas, hot water or steam rises to the surface. In other places, cold water can be pumped down to the hot rocks, and returns as hot water. The heat energy is used to drive turbines connected to electricity generators.

>> practice questions

1 Outline three ways that the energy in moving water can be used to generate electricity.

2 Suggest why solar cells are used to supply electricity to satellites.

3 Explain why geothermal energy is sometimes described as only partly renewable.

Resources compared

- Fossil fuels, nuclear fuels and renewable energy sources have different advantages and disadvantages for generating electricity.

- Factors to take into account include building costs, start-up times, reliability, the relative cost of the electricity produced, and the place where it is needed.

A Building costs and fuel costs

1 **key fact** Different types of power station have different building costs. This affects the cost of the electricity produced during the lifetime of the power station.

2 In general, large and complex power stations cost the most to build. Nuclear power stations are very complex machines. They are very expensive to build but nuclear fuels are relatively cheap. Hydroelectric power schemes and tidal barrages are expensive but do not need any fuel.

3 Gas and oil are expensive fuels compared to coal. Wind turbines and solar cells do not need any fuel.

B Start-up times

1 **key fact** Different types of power station have different start-up times. This affects the way in which the power station is used.

2 Hydroelectricity stations can be started almost immediately.
Wind turbines start to work as soon as the wind reaches a certain speed.
Solar cells produce electricity as long as light falls on them.

type of power station	start-up time
gas-fired	shortest
oil-fired	
coal-fired	
nuclear	longest

The main types of power stations have different start-up times.

C Reliability

1 **key fact** Some types of power station are more reliable than others.

2 Tidal barrages are reliable. They generate electricity as the tide comes in, and this happens twice a day. Some designs also work as the tide goes out.

3 Hydroelectric power stations are reliable. Their dams store water so that electricity can be generated all year round, and not just when a river is flowing quickly.

4 Wind turbines and solar cells are less reliable. Wind turbines only work when it is windy, and solar cells only work during the day.

exam tip >>

Be prepared to use information in the exam to discuss the advantages and disadvantages of using different energy sources to generate electricity.

>> practice questions

1 Extra electricity is needed at peak times of the day. Suggest why oil-fired and gas-fired power stations usually provide this electricity.

2 'Base load electricity' is generated all the time, whatever the demand on the National Grid. Suggest why nuclear power stations and coal-fired power stations usually provide this electricity.

3 Explain why electricity is supplied to the National Grid from many different types of power station, rather than just one type.

The electromagnetic spectrum

- Electromagnetic radiation travels as waves.

- Waves can be described by their wave speed, frequency and wavelength.

- The electromagnetic spectrum can be grouped into different types, such as visible light and radio waves.

- All types of electromagnetic radiation travel at the same speed in space.

A Waves

1 Waves along a rope, water waves and light waves are all examples of transverse waves. The vibrations are at right angles to the direction of travel.

2 The amplitude of a wave is its maximum disturbance from its undisturbed position.

3 **key fact** Waves move energy from one place to another.

4 **key fact** The wavelength of a wave is the distance between a point on one wave and the same point on the next wave. The wavelength is measured in metres, m.

5 **key fact** The frequency of a wave is the number of waves that pass a certain point each second. Frequency is measured in hertz, Hz. 1 Hz is one wave per second.

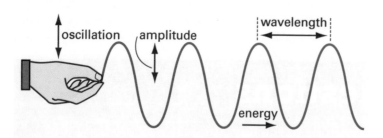

B Wave speed

① key fact Electromagnetic waves can be described by this equation:

$$\text{wave speed (metre/second, m/s)} = \text{frequency (hertz, Hz)} \times \text{wavelength (metre, m)}$$

② All electromagnetic waves travel at the same speed in a vacuum, such as space. This is extremely fast, about 300 000 000 m/s, but you don't need to remember the number.

③ *Example* A radio station broadcasts at 100 MHz (100 000 000 Hz). What is the wavelength of its radio waves?

Answer Re-arrange the equation:

wavelength = wave speed ÷ frequency

wavelength = 300 000 000 ÷ 100 000 000 = 3 m

C The electromagnetic spectrum

① key fact The electromagnetic spectrum is a continuous range of waves, from waves with a very short wavelength to those with a very long wavelength.

exam tip >>

Make sure you know the order of the different types of radiation in the electromagnetic spectrum.

② key fact Gamma rays have the shortest and radio waves have the longest wavelengths.

③ key fact The longer the wavelength, the lower the frequency. So gamma rays have the highest frequency and radio waves have the lowest frequency.

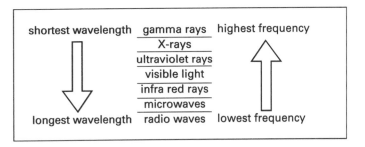

shortest wavelength — gamma rays — highest frequency
X-rays
ultraviolet rays
visible light
infra red rays
microwaves
longest wavelength — radio waves — lowest frequency

>> practice questions

1 Write the equation that links wave speed, frequency and wavelength.

2 A wireless computer network operates at 2.4 GHz (2 400 000 000 Hz). If the wavelength is 0.125 m, calculate the wave speed.

3 Which type of electromagnetic radiation has the higher frequency, ultraviolet rays or infra red rays?

Some uses of electromagnetic radiation

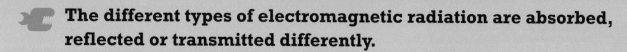 The different types of electromagnetic radiation are absorbed, reflected or transmitted differently.

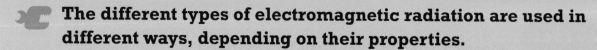 The different types of electromagnetic radiation are used in different ways, depending on their properties.

A Gamma radiation

1. Gamma radiation cannot be seen or felt. It mostly passes through skin and soft tissue, but some of it is absorbed by cells.

2. Gamma radiation is used to:
   - kill cancer cells
   - kill harmful bacteria in food
   - sterilise surgical instruments.

B X-rays

1. X-rays cannot be seen or felt. They mostly pass through skin and soft tissue, but they do not easily pass through bone or metal.

2. X-rays are used to:
   - produce photographs of bones to check for damage such as fractures
   - check metal components and welds for cracks or other damage.

C Ultraviolet radiation

1. We cannot see or feel ultraviolet radiation, but our skin responds to it by turning darker.

2. Ultraviolet radiation is used in:
   - sun beds
   - security pens
   - fluorescent lights.

D Infra red radiation

1. Infra red radiation is absorbed by the skin and we feel it as heat.

2. Infra red radiation is used in:

- night vision cameras
- television remote controls
- heaters, grills and toasters
- optical fibre communications.

remember >>

Thermal energy is transferred through empty space by infra red radiation.

E Microwaves and radio waves

1. Microwaves with certain wavelengths are absorbed by water molecules and can be used for cooking:

- water, fats and sugars easily absorb microwaves
- their molecules vibrate more vigorously, making the food in the oven hot.

2. Microwave radiation is also used in mobile phone and satellite communications.

3. Radio waves are used to transmit television and radio programmes.

exam tip >>

Microwaves and radio waves are absorbed by metal antennae, creating an alternating electrical current with the same frequency as the radiation.

>> practice questions

1. How do our bodies detect infra red radiation if we cannot see it?

2. Outline how a microwave oven works.

3. Explain why X-rays can be used to check machinery for cracks without taking it apart.

Communication signals

- Digital signals have advantages over analogue signals.
- Electromagnetic radiation can be used for communication.
- Visible light and infra red rays travel through optical fibres.
- Microwaves are used for satellite communications and mobile phones.

A Signals

① **key fact** Communication signals may be analogue or digital.

② **key fact** Analogue signals vary continuously.

The shape of an analogue signal matches the original wave.

③ **key fact** Digital signals usually only have two values, 0 or 1. This means that they are more easily processed by computers.

④ Signals become weaker as they travel. They also pick up random extra signals called noise. Noise causes hiss on radio programmes, and it causes internet connections to slow down as the modem tries to compensate.

⑤ **key fact** Digital signals are less affected by noise than analogue signals.

B Optical fibres

① An optical fibre is a thin strand of glass with a protective coating. Light repeatedly reflects off the fibre's inner surface. In this way, light going in at one end emerges at the other end, even when the fibre is bent.

② **key fact** Infra red rays and visible light are used to send signals through optical fibres.

An optical fibre carries more information, and over greater distances, than an ordinary cable of the same thickness.

Light can follow a curved path through an optical fibre.

C Radio waves and microwaves

Radio waves are used to transmit television and radio programmes. Long wavelength radio waves are reflected by the upper atmosphere. This lets them reach receivers over the horizon.

key fact Microwaves are used to send information to and from satellites, and in mobile phone networks.

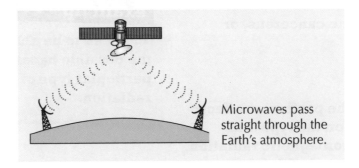

Microwaves pass straight through the Earth's atmosphere.

People are advised to keep their exposure to microwaves from mobile phones to a minimum.

>> practice questions

1 Outline the differences between analogue signals and digital signals.

2 Explain why light can travel from one end to the other through an optical fibre.

3 Why are microwaves used to transmit information to and from satellites?

Hazards of electromagnetic radiation

- Electromagnetic radiation can affect living cells.
- The effects on cells depend on the type and amount of electro-magnetic radiation.

A Absorption of electromagnetic radiation

1 Exposure to very bright visible light can damage our eyesight. People who work with bright light, especially laser light, need to wear eye protection.

2 **key fact** Some types of electromagnetic radiation mostly pass through soft tissue, such as skin and muscle, without being absorbed.

3 **key fact** Some types of electromagnetic radiation are absorbed by soft tissue. This can cause:

- the tissue to heat up
- cells to become cancerous, or
- death of cells.

exam tip >>

You need to be able to evaluate the possible hazards of using a particular type of electromagnetic radiation.

4 **key fact** The particular effects on cells depend on the type and amount of electromagnetic radiation. For a given amplitude, high frequency radiation transfers more energy than low frequency radiation.

B Gamma radiation and X-rays

1 Gamma radiation and X-rays are very penetrating. Exposure to these types of electromagnetic radiation increases the risk of cancer.

2 Lead absorbs gamma radiation and X-rays. People who work with these types of radiation can minimise their exposure using lead shielding. For example, hospital radiographers may leave the room or stand behind a lead screen. They and their patients may wear lead aprons.

3 Radiotherapy treatment for cancer may use several weak beams of gamma radiation or X-rays that focus on the tumour. This way, healthy cells do not receive too much radiation and the tumour cells receive enough to kill them.

C Ultraviolet light

1 Ultraviolet radiation is found naturally in sunlight. It causes skin cancer.

2 Our skin responds to exposure to ultraviolet light by turning darker. This reduces the amount that reaches deeper skin tissues.

3 Exposure to ultraviolet light can be reduced by avoiding the sun and by using sunscreen.

D Microwaves

1 Microwave ovens are designed to prevent the user being exposed to microwave radiation. Safety features include:

- a metal mesh in the door to prevent microwaves escaping

- a safety lock that stops the oven working if the door is open.

2 There is concern that exposure to microwave radiation from mobile phones may be harmful, but the scientific evidence is not conclusive.

3 All mobile phones sold in the UK must produce SARs (Specific Absorption Rates) below a certain level. Users are advised to reduce call times and increase the distance of the phone from the head, for example by using a hands-free kit.

exam tip >>

You may be asked to evaluate information you are given about different ways to reduce exposure to electromagnetic radiation.

>> practice questions

1 Explain why a hospital radiographer may wear a lead apron.

2 Suggest why people are advised to wear hats and finely woven, loose fitting clothes when they go outside in the summer.

3 Describe two ways in which exposure to radiation from microwave ovens is prevented.

Atomic radiation

⚙ Isotopes are atoms of an element with the same number of protons but different numbers of neutrons.

⚙ Radioactive substances give out radiation from the nuclei of their atoms.

A Atoms and isotopes

① **key fact** Atoms have a small central nucleus made of particles called protons and neutrons. Even smaller particles called electrons surround the nucleus.

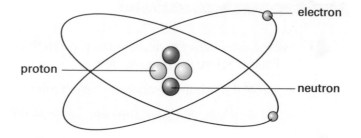

② **key fact** Atoms of an element all have the same number of protons. Atoms of different elements have different numbers of protons.

③ **key fact** Isotopes are atoms of an element that have the same number of protons but a different number of neutrons. For example, carbon-12 and carbon-14 are isotopes of carbon. Their atoms have six protons, but carbon-12 atoms have six neutrons and carbon-14 atoms have eight neutrons.

④ Some isotopes are radioactive. They are often called radioisotopes.

B Radioactivity

① Atomic radiation consists of particles or electromagnetic waves given off by the nuclei of atoms.

② **key fact** Substances that give off radiation are radioactive.

③ **key fact** Radioactive substances give off radiation all the time, whatever is done to them. Chemical reactions, heating or cooling have no effect on the radiation.

C Types of radiation

1 key fact There are three main types of atomic radiation.

- Alpha radiation consists of high energy particles from the nucleus. Each alpha particle is made from two protons and two neutrons, joined together.

- Beta radiation also consists of high energy particles from the nucleus. Each beta particle is an electron, formed when a neutron breaks down.

- Gamma radiation is high energy electromagnetic radiation from the nucleus.

2 They are called ionising radiation. This is because, when they collide with an atom, they can knock an electron out of the atom. The atom becomes an electrically charged particle called an ion.

D Properties of radiation

1 key fact Alpha radiation and beta radiation are deflected from moving in straight lines by electric fields and magnetic fields. Gamma radiation is not affected.

2 key fact Alpha radiation is the most ionising type of radiation and gamma radiation is the least ionising type.

radiation	distance travelled in air	stopped by		
		paper	thin aluminium	thick lead
alpha radiation	a few cm	✔	✔	✔
beta radiation	a few tens of cm		✔	✔
gamma radiation	many metres			✔

>> practice questions

1 What are isotopes?

2 What types of radiation consists of particles?

3 Outline two differences in the properties of alpha, beta and gamma radiation.

- Radioactive substances give off less radiation as time goes by.

- Half-life is how long it takes for the number of nuclei of a radioactive isotope in a sample to halve.

- Different radioactive isotopes have different uses depending on factors such as their penetrating power, their ionising power, and their half-life.

A Half-life

1. Radioactive nuclei are unstable. They may decay at any time, releasing radiation. This radioactive decay helps the nucleus to become more stable.

2. **key fact** Radioactive decay is a random process. You cannot predict when any individual nucleus will decay. But you can say that, after a certain amount of time, half the nuclei in a sample of the material will have decayed. This is the idea behind half-life.

3. **key fact** The half-life of a radioactive isotope is defined as:

   - **the time taken for half the nuclei of that isotope in a sample to decay, or**

   - **the time taken for the count rate from a sample to fall to half the starting rate.**

4. Different radioactive isotopes have different half-lives. They vary from fractions of a second to millions of years.

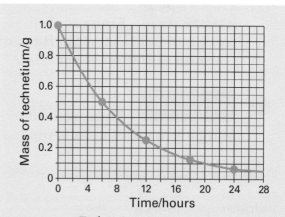

Technetium-99 is a gamma emitter. The graph shows how a sample of it decays. Technetium-99 has a half-life of six hours. So, if there is one gram at the start, after six hours this will have decayed to 0.5 g. After another six hours there will only be 0.25 g left.

remember >>

The shape of the graph will be the same for all radioactive isotopes, it is just the axis labels that will change.

exam tip >>

The half-life will be the same wherever you measure it off the graph. Try to pick a sensible place on the graph.

B Uses of radiation

1 **key fact** Some sources of radiation are better than others for particular jobs. When choosing the best one, take into account:

- the penetrating power
- the ionising power
- the half-life.

2 Alpha radiation does not travel very far, but it is very ionising. Sources of alpha radiation can damage living cells if swallowed. Gamma emitters are a better choice for medical imaging. Gamma radiation is the least ionising, and it will travel right out of the body to the detectors.

3 Radioactive tracers are used to follow chemicals in the environment or in the body. It is important that their half-life is long enough for the scientists or doctors to collect results, but not so long that they persist for a long time afterwards.

exam tip >>

Be prepared to analyse information about different radioactive isotopes to choose the best one for a particular job.

>> practice questions

1 State one definition of the term 'half-life'.

2 Iodine-125 has a half-life of 60 days. Starting with 8 g of iodine-125, how much will be left after 240 days?

3 Explain why technetium-99 is suitable for use in medical imaging. Give two reasons in your answer.

Observing the universe

- Telescopes are used on Earth and in space to observe the universe.

- When a wave source moves towards or away from an observer, its observed wavelength and frequency changes.

- Light from distant galaxies is shifted towards the red end of the spectrum.

- Red-shift data provides evidence about the nature of the universe.

A Optical telescopes

1. Different types of telescope can observe the universe at different frequencies of electromagnetic radiation.

2. Optical telescopes are sensitive to visible light. Amateur astronomers use small telescopes, but very large ones are used by professional astronomers.

3. Ground-based optical telescopes can be used only at night and if the weather is good.

B Other telescopes

1. Objects in the universe do not just give off visible light. They also give off other electromagnetic radiation including radio waves, microwaves, X-rays and gamma rays.

2. Some telescopes can detect this radiation. They can be used in the daytime as well as at night, and they can be used in poor weather, because the radiation is not blocked by clouds.

3. Telescopes that detect X-rays and gamma rays produce very detailed images. This is because the wavelength of the electromagnetic radiation they detect is very small.

4. Space-based telescopes can be used all the time, but are very expensive and difficult to maintain.

> **exam tip >>**
>
> Be prepared to use information in the exam to discuss the advantages and disadvantages of different types of telescope.

C The nature of the universe

1 Scientists have gathered a lot of evidence about the nature of the universe.

2 **key fact** **The accepted model of the universe is the Big Bang theory. It states that about 13.6 billion years ago all the matter in the universe was concentrated into a single very small point. This enlarged rapidly in a hot explosion. The universe is still expanding today.**

D Red-shift

1 The wavelength and frequency of a source of waves changes if it moves towards or away from you. This is why the pitch of an ambulance siren rises and falls as it passes you.

2 **key fact** **The light from galaxies is red-shifted because the observed wavelength of the light has increased. This is evidence that galaxies are moving away from us.**

3 **key fact** **The further away the galaxy, the more its light is red-shifted. This means that the more distant galaxies are moving away more quickly than nearby ones.**

4 **key fact** **This is evidence for the Big Bang: in an explosion, the fastest moving objects end up furthest away.**

> **exam tip >>**
> **You do not need to know the order of the planets for your exam.**

>> practice questions

1 Give two advantages of radio telescopes over optical telescopes.

2 Outline the Big Bang theory of the origin of the universe.

3 Explain how red-shift evidence supports the Big Bang theory.

Exam questions

Specimen QUESTION ONE

This question is about white blood cells and how they help to defend the body against pathogens.

Match words, **A**, **B**, **C** and **D**, with the numbers **1 – 4** in the sentences.

A   antibodies

B   antigens

C   antitoxins

D   ingest

Pathogens have chemicals called … **1** … on their surface. White blood cells produce … **2** … and these stick to the pathogens.

White blood cells … **3** … pathogens and destroy them with reactive chemicals.

White blood cells also produce … **4** … that counteract toxins. **4 marks**

Specimen QUESTION TWO

The diagram shows how nerve impulses travel in a reflex action.

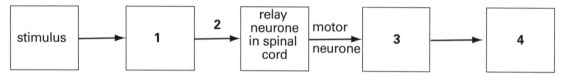

Match words, **A**, **B**, **C** and **D**, with the numbers **1 – 4** on the flow chart.

A   effector

B   receptor

C   response

D   sensory neurone **4 marks**

Specimen QUESTION THREE

The conditions in the body must be controlled.

Match words, **A**, **B**, **C** and **D**, with the numbers **1 – 4** in the sentences.

A   ions

B   lungs

C   sugar

D   urine

Water leaves the body from the … **1** … when we breathe out. Water and … **2** … leave the body from the kidneys in … **3** … The blood … **4** … level must be controlled to provide cells with a constant source of energy. **4 marks**

Specimen QUESTION FOUR

This question is about tobacco smoke and alcohol, and the harm they can do to the body.

Match substances, **A**, **B**, **C** and **D**, with the features **1 – 4** in the table.

A alcohol

B carbon monoxide

C nicotine

D tar

	feature
1	the addictive substance in tobacco smoke
2	reduces the ability of red blood cells to carry oxygen
3	causes lung cancer
4	causes a lack of self control

4 marks

> All the remaining questions have four parts.
> Choose only one answer in each part, but you can use the same number more than once.

Specimen QUESTION FIVE

5A Which row in the table shows the correct source of each hormone?

	oestrogen	FSH	LH
1	ovaries	ovaries	pituitary
2	pituitary	ovaries	ovaries
3	ovaries	pituitary	ovaries
4	ovaries	pituitary	pituitary

5B Different hormones have different effects.

1 Oestrogen causes eggs to mature.

2 FSH causes eggs to mature.

3 FSH inhibits the production of oestrogen.

4 Oestrogen is useful as a fertility drug.

5C Information passes across a synapse …

1 by the diffusion of a chemical.

2 by nerve impulses crossing the gap.

3 by nerve impulses through a motor neurone.

4 by hormones produced by the pituitary.

5D Your metabolic rate varies with the amount of activity you do.

1 The less exercise you take and the warmer it is, the less food you need.

2 The less exercise you take and the warmer it is, the more food you need.

3 The more exercise you take and the warmer it is, the more food you need.

4 The more exercise you take and the warmer it is, the less food you need.

4 marks

Specimen QUESTION SIX

The graph shows the changes in the concentration of certain antibodies in the blood following vaccination.

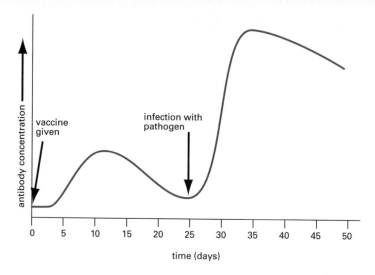

Graph of primary and secondary immune response.

6A How long after infection by the pathogen is the maximum antibody concentration reached?

1 10 days

2 12 days

3 25 days

4 37 days

6B Compared to the first immune response, the second immune response is …

1 faster and larger.

2 faster and smaller.

3 slower and smaller.

4 slower and larger.

6C A vaccine may contain …

1 a dead form of the disease.

2 an inactive form of the pathogen.

3 red blood cells that produce antibodies.

4 a mixture of antibiotics to kill pathogens.

6D The data on antibody concentration are presented as a line graph. This is because …

1 the dependent variable is categoric and the independent variable is continuous.

2 the dependent variable is continuous and the independent variable is categoric.

3 both variables are categoric.

4 both variables are continuous.   **4 marks**

Unit B1b Evolution and Environment

Specimen QUESTION ONE

The question is about competition between living things.

Match words, **A**, **B**, **C** and **D**, with the numbers **1 – 4** in the sentences.

A   food

B   light

C   nutrients

D   territory

Animals defend their … **1** … and compete with each other for … **2** …

Plants compete with each other for … **3** … from the soil and … **4** … **4 marks**

Specimen QUESTION TWO

The polar bear lives in the Arctic, where it is cold. Polar bears have adaptations that let them survive in the conditions found in the Arctic.

Match adaptations, **A**, **B**, **C** and **D**, with the numbers **1 – 4** in the table.

A   broad feet with claws

B   good sense of smell

C   thick layer of fat

D   white fur

	reason for adaptation
1	ice and snow is slippery
2	camouflage so prey cannot see them
3	to stay warm in the cold
4	to find prey for food

4 marks

Specimen QUESTION THREE

Human activities can affect the Earth's climate.

Match words, **A**, **B**, **C** and **D**, with the numbers **1 – 4** in the sentences.

A   carbon dioxide

B   fossil fuels

C   methane

D   thermal

Greenhouse gases absorb … **1** … energy radiated by the Earth.

The use of … **2** … increases the amount of … **3** … in the air.

Cattle and rice paddy fields increase the amount of … **4** … in the air. **4 marks**

Specimen QUESTION FOUR

This question is about the cloning of animals.

Match words, **A**, **B**, **C** and **D**, with the numbers **1 – 4** in the sentences.

A   embryo

B   fusion

C   unfertilised egg

D   womb

The nucleus of an … **1** … cell is removed.

The nucleus from an … **2** … cell is then injected into the cell without a nucleus.

The egg cell is implanted into the … **3** … of a host mother, where it develops.

This is called … **4** … cell cloning. **4 marks**

> All the remaining questions have four parts.
> Choose only one answer in each part, but you can use the same number more than once.

Specimen QUESTION FIVE

The theory of evolution explains how the different species of living things arose on Earth.

5A One reason why Darwin's theory of evolution by natural selection was only gradually accepted by most scientists was that …

1   Darwin had no evidence.

2   Lamarck's theory of evolution was more accurate.

3   Darwin's theory conflicted with religious views.

4   most scientists refused to analyse Darwin's theory.

5C New forms of genes can appear because of …

1   variation between individuals in a species.

2   sexual reproduction.

3   mutation.

4   cloning.

5B The dinosaurs became extinct millions of years ago. Which is the least likely reason for this?

1   New competitors evolved.

2   New predators evolved.

3   The climate changed.

4   People hunted them.

5D   The theory of evolution states that all species of living things have evolved from simple life-forms which first developed about …

1   three thousand years ago.

2   three million years ago.

3   three hundred million years ago.

4   three billion years ago.   **4 marks**

Specimen QUESTION SIX

Reducing the area of forest is called deforestation. The graph shows the amount of deforestation in the Amazon each year between 1988 and 2007.

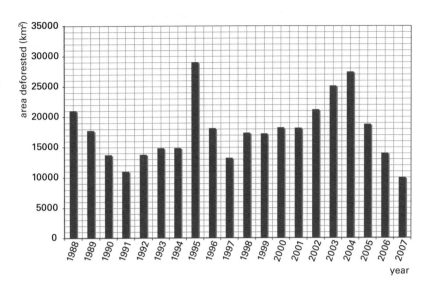

6A The data are presented as a bar chart. This is because …

1 the dependent variable is categoric and the independent variable is continuous.

2 the dependent variable is continuous and the independent variable is categoric.

3 both variables are categoric.

4 both variables are continuous.

6B The mean amount of deforestation each year is about …

1 10 000 km²

2 12 500 km²

3 17 500 km²

4 29 000 km²

6C Which statement is a valid conclusion from these data?

1 Deforestation is caused by human activity.

2 Deforestation is caused by natural events.

3 The amount of deforestation is almost the same each year.

4 The area of forest in the Amazon decreased in each year of the study.

6D Scientists were interested to see why deforestation was particularly high in 1995. The best way to gather evidence about this would be to …

1 analyse satellite photographs of the Amazon taken in 1994 and 1995.

2 analyse data from 2004, as deforestation was also high then.

3 interview some local people for their ideas about the forest.

4 send scientists to all the deforested areas.

4 marks

Specimen QUESTION ONE

This question is about atoms.

Match words, **A**, **B**, **C** and **D**, with the numbers **1 – 4** in the sentences.

A   bonds

B   electrons

C   element

D   nucleus

Atoms have a central … **1** … surrounded by … **2** … A substance made from one sort of atom
is called an … **3** … The atoms in a compound are held together by chemical … **4** …   **4 marks**

Specimen QUESTION TWO

Limestone is a useful material.

The flow chart shows how some of its products
are made.

Match adaptations, **A**, **B**, **C** and **D**, with the
numbers **1 – 4** in the flow chart.

A   carbon dioxide

B   slaked lime

C   heat in a kiln

D   add water   **4 marks**

```
┌───────────┐   1   ┌───────────┐   3   ┌───────────┐
│ limestone │ ────▶ │ quicklime │ ────▶ │     4     │
└───────────┘       └───────────┘       └───────────┘
                          +
                    ┌───────────┐
                    │     2     │
                    └───────────┘
```

Specimen QUESTION THREE

Natural gas and many other hydrocarbons are useful as fuels.
They produce many substances when they burn.

Match the substances, **A**, **B**, **C** and **D**, with the numbers **1 – 4** in the table.

A   carbon dioxide

B   natural gas

C   soot

D   sulfur dioxide

	what we can say about the substance
1	causes global warming
2	causes global dimming
3	causes acid rain
4	contains carbon and hydrogen

4 marks

Specimen QUESTION FOUR

This question is about metals and how they are extracted.

Match words, **A**, **B**, **C** and **D**, with the numbers **1 – 4** in the table.

A   copper

B   gold

C   iron

D   titanium

	what we can say about the metal
1	It is usually converted into steel.
2	It is expensive to extract because a lot of energy is needed.
3	It has properties that make it useful for electrical wires and plumbing.
4	It is an unreactive metal that can be found in the ground in its native state.

4 marks

> All the remaining questions have four parts.
> Choose only one answer in each part, but you can use the same number more than once.

Specimen QUESTION FIVE

Crude oil can be separated into fractions using fractional distillation.

5A Fractional distillation separates the hydrocarbons in crude oil according to their …

1   density.

2   colour.

3   boiling point.

4   chemical reactivity.

5B   Which statement is correct?

1   The fraction with the lowest boiling point and lowest viscosity is found at the top of the column.

2   The fraction with the highest boiling point and lowest viscosity is found at the top of the column.

3   The fraction with the highest boiling point and highest viscosity is found at the top of the column.

4   The fraction with the lowest boiling point and highest viscosity is found at the top of the column.

5C   The chemical formula of ethane is C_2H_6 and the formula of propane is C_3H_8.
The formula of pentane is:

1 C_5H_8

2 C_5H_{10}

3 C_5H_{12}

4 C_5H_{14}

5D   Methane burns completely in air to produce carbon dioxide and water vapour.
The correct equation for this reaction is:

1 $CH_4 + O_2 \rightarrow CO_2 + 2H_2O$

2 $CH_4 + 2O_2 \rightarrow CO_2 + 2H_2O$

3 $CH_4 + O_2 \rightarrow CO + 2H_2O$

4 $CH_4 + 2O_2 \rightarrow 2CO + 2H_2O$ **4 marks**

Specimen QUESTION SIX

Read the information in the box about recycling aluminium.

It takes 14 kWh of electricity to make 1 kg of aluminium from aluminium ore. Making aluminium by recycling uses just 5% of the electricity needed to make aluminium from aluminium ore. Every time 1 kg of aluminium is recycled, 8 kg of aluminium ore is saved and 4 kg of waste is not produced.

6A The amount of electricity saved by recycling 1 kg of aluminium is:

1 0.7 kWh

2 7 kWh

3 13.3 kWh

4 13.5 kWh

6C Recycling aluminium uses less electricity than making aluminium from aluminium oxide because …

1 the aluminium oxide has aleady been reduced.

2 most aluminium is mixed with other metals to make alloys.

3 aluminium has a low density and resists corrosion.

4 aluminium is a transition metal.

6B An aluminium can weighs about 15 g. How much waste is saved by recycling 80 cans, the average number used by each person in the UK each year?

1 60 g

2 320 g

3 1200 g

4 4800 g

6D Aluminium is more expensive to produce than iron. This is because …

1 there is more iron ore in the Earth's crust than aluminium ore.

2 iron is more reactive than aluminium.

3 iron can be produced by reaction with carbon, but aluminium cannot.

4 pure iron is very soft.　　　　**4 marks**

Unit C1b Oils, Earth and Atmosphere

Specimen QUESTION ONE

The atmospheres of Mars and Earth are mixtures of gases.

Match gases, **A**, **B**, **C** and **D**, with the numbers **1 – 4** in the table.

A   carbon dioxide

B   helium

C   nitrogen

D   oxygen

	what we can say about the gas
1	It is the main gas in the atmosphere of Mars.
2	It is produced by photosynthesis in plants.
3	It is the main gas in the Earth's atmosphere.
4	It is one of the noble gases.

4 marks

Specimen QUESTION TWO

This question is about the Earth's tectonic plates.

Match words, **A**, **B**, **C** and **D**, with the numbers **1 – 4** in the sentences.

A   convection

B   crust

C   heat

D   mantle

The Earth's … **1** … and upper part of the … **2** … are cracked into large pieces called tectonic plates.

The plates move because of … **3** … currents driven by … **4** … from natural radioactive processes inside the Earth. **4 marks**

Specimen QUESTION THREE

Hydrocarbon molecules in oil fractions can be cracked at the oil refinery.

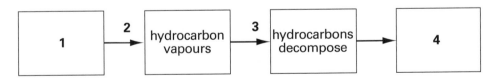

Match words, **A**, **B**, **C** and **D**, with the numbers **1 – 4** in the flow chart.

A   hydrocarbons are heated

B   small hydrocarbon molecules

C   large hydrocarbon molecules

D   vapours are passed over a hot catalyst **4 marks**

Specimen QUESTION FOUR

This question is about vegetable oils and their reactions.

Match words, **A**, **B**, **C** and **D**, with the numbers **1 – 4** in the sentences.

A   emulsions

B   nickel

C   iodine

D   hydrogen

The presence of double carbon–carbon bonds can be detected because of a colour change
when … **1** … is mixed with vegetable oils.

Unsaturated oils can be hardened to make them suitable for cakes and margarine by reacting them with …
2 … at about 60 °C with a catalyst of … **3** …

Vegetable oils form … **4** … when shaken with water. **4 marks**

> All the remaining questions have four parts.
> Choose only one answer in each part, but you can use the same number more than once.

Specimen QUESTION FIVE

Hydrocarbons called alkenes are produced by cracking.

5A What is the general formula for an alkene?

1   C_nH_{2n+2}

2   C_nH_{n+2}

3    $C_{2n}H_n$

4   C_nH_{2n}

5C The chemical formula of ethene is C_2H_4 and the formula of propene is C_3H_6.
The formula of pentene is:

1   C_5H_8

2   C_5H_{10}

3   C_5H_{12}

4   C_5H_{14}

5B Why are alkenes described as unsaturated?

1   They contain double carbon–carbon bonds.

2   They are hydrocarbons.

3   They contain single carbon–carbon bonds.

4   They react with oxygen.

5D Ethanol is useful as a fuel. It can be produced by the fermentation of sugar, or by the reaction of ethene with steam. One advantage of producing ethanol by fermentation is that …

1   it uses non-renewable resources.

2   sugar is a renewable resource.

3   it takes place in the presence of a catalyst.

4   ethanol is the only product of the reaction of ethene with steam.
4 marks

Specimen QUESTION SIX

Artificial colours in food can be detected and identified by chromatography.

The different substances can be identified by their Rf (retention factor) value:

$$R_f = \frac{\text{distance travelled by the substance}}{\text{distance travelled by the solvent}}$$

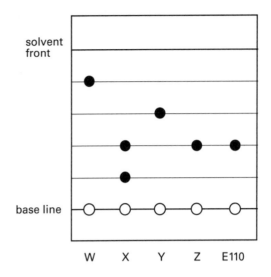

6A In the chromatogram, what is the Rf value of E110?

1   0.2

2   0.4

3   0.6

4   2.5

6B From the information in the chromatogram, what can be said about the substances W, X, Y and Z?

1   X and Z are the same substance.

2   X is a mixture of W and Z.

3   Z is found in X but not in Y.

4   Z is more soluble in the solvent than W.

6C E110 is also called sunset yellow. The chromatography results show that …

1   Z must be yellow.

2   X and Z must contain E110.

3   Z could be sunset yellow.

4   X is likely to be orange coloured.

6D Food additives may be given E numbers to show that …

1   they are artificial.

2   they could cause allergies in some people.

3   they improve the shelf life of the food.

4   they are permitted for use in food.

4 marks

105

Unit P1a Energy and Electricity

Specimen QUESTION ONE

Televisions transform energy.

Match types of energy, **A**, **B**, **C** and **D**, with the numbers **1 – 4** in the sentences.

A   light

B   electrical

C   thermal

D   sound

The energy input to the television is … **1** … energy. The useful type of energy output by the screen is … **2** … energy. The useful type of energy output by the loudspeakers is … **3** … energy. The type of energy not usefully output by the television is … **4** … energy. **4 marks**

Specimen QUESTION TWO

The use of renewable energy resources to generate electricity reduces the need to use fossil fuels.
This question is about advantages and disadvantages of renewable energy resources.

	feature
1	convenient for supplying electricity to small portable devices
2	farmland is flooded
3	may be seen from a long distance and so spoil the view
4	mostly only available in volcanic areas

Match the words, **A**, **B**, **C** and **D**, with the numbers **1 – 4** in the table.

A   geothermal energy stations

B   solar cells

C   wind farms

D   hydroelectric power schemes   **4 marks**

Specimen QUESTION THREE

The flow chart shows the main events in a conventional power station.

Match the words, **A**, **B**, **C** and **D**, with the numbers **1 – 4** in the sentences.

A   electrical

B   kinetic

C   steam

D   thermal

The burning fuel releases … **1** … energy.

The turbine spins because … **2** … pushes against it.

The turbine transfers … **3** … energy to the generator.

The generator produces … **4** … energy. **4 marks**

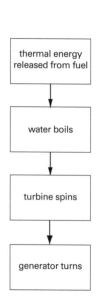

Specimen QUESTION FOUR

A vacuum flask maintains the temperature of drinks longer than ordinary containers.

It has several features to help reduce the transfer of thermal energy to and from it.

Match features, **A**, **B**, **C** and **D**, with the numbers **1 – 4** in the table.

A   silvery walls

B   stopper on the flask

C   thin glass walls

D   vacuum between the glass walls

	reason
1	reduces heat losses from the top by convection
2	reduces heat losses from the sides by convection
3	reduces absorption and emission of infra red radiation
4	reduces heat losses through the wall by conduction

4 marks

> All the remaining questions have four parts.
> Choose only one answer in each part, but you can use the same number more than once.

Specimen QUESTION FIVE

Modern electrical appliances such as fridges and dishwashers are given 'efficiency ratings'.
G is the least efficient and A is the most efficient.

5A Compared to a G rated dishwasher, an
A rated dishwasher …

1   usefully transforms a bigger proportion of the
electrical energy supplied to it.

2   usefully transforms a smaller proportion of the
electrical energy supplied to it.

3   uses less water.

4   dries the dishes for you.

5B The power of a certain dishwasher is listed as
1250 W. This is the same as:

1   125 kW

2   12.5 kW

3   1.25 kW

4   0.125 kW

5C An electric light bulb releases 4 J of light energy
when it is supplied with 16 J of electrical energy.
What is its efficiency?

1   0.25

2   0.40

3   0.75

4   4.00

5D How much electrical energy, in kWh, is
transferred when a 0.2 kW television set
is used for 30 minutes?

1   0.1

2   2.5

3   6.0

4   150

4 marks

Specimen QUESTION SIX

Different energy resources can be used to generate electricity. Scientists and politicians must carefully consider the advantages and disadvantages of each energy resource. They must ensure that the UK continues to generate the electricity it needs, while minimising harm to the environment.

6A A typical wind turbine needs 1.2 million kWh of electricity in its manufacture. It generates 96 million kWh of electricity in its 20 year lifetime. How long is needed for the turbine to generate more electricity than was needed to make it?

1   3 months

2   4 months

3   3 years

4   6.7 years

6C Which of the following are non-renewable energy resources?

1   natural gas and tidal power

2   oil and uranium

3   wind power and uranium

4   wave energy and coal

6B Which of the following is a disadvantage of nuclear power stations?

1   They produce almost no greenhouse gases while in use.

2   They run continuously so they supply 'base load' electricity.

3   Nuclear waste is difficult to handle and store safely.

4   Nuclear fuels release heat in a nuclear reactor.

6D Which of the following energy resources does not ultimately rely on energy from the Sun?

1   solar power

2   hydroelectric power

3   wind power

4   tidal power

4 marks

Unit P1b Radiation and the Universe

Specimen QUESTION ONE

Different types of electromagnetic radiation have different uses.

Match the types of electromagnetic radiation, **A**, **B**, **C** and **D**, with the numbers **1 – 4** in the table.

A   X-rays

B   infra red rays

C   microwaves

D   ultraviolet rays

	use
1	making medical images
2	mobile telephone communications
3	tanning beds
4	television remote controls

4 marks

Specimen QUESTION TWO

The different types of electromagnetic radiation can be put in order of decreasing frequency.

Match words, **A**, **B**, **C** and **D**, with the numbers **1 – 4** in the table.

A   infra red light

B   visible light

C   radio waves

D   X-rays

gamma	1	ultraviolet	2	3	microwaves	4

decreasing frequency

4 marks

Specimen QUESTION THREE

Electromagnetic radiation can harm us if it is used incorrectly.

Match words, **A**, **B**, **C** and **D**, with the numbers **1 – 4** in the sentences.

A   infra red radiation

B   X-rays

C   ultraviolet light

D   microwave radiation

Exposure to … **1** … can cause skin cancer. Cells are damaged when … **2** … heats up the water inside them.

There is a risk from exposure to … **3** … of causing mutations in unborn babies. … **4** … is felt as heat and can cause burns. **4 marks**

Specimen QUESTION FOUR

Telescopes are used to observe the universe and to provide scientists with evidence about it.

Match words, **A**, **B**, **C** and **D**, with the numbers **1 – 4** in the sentences.

A the universe

B galaxies

C point

D Earth

Telescopes on … **1** … that detect radio waves can work all the time.

Light from distant … **2** … is red-shifted.

The observed red-shift is evidence that … **3** … began as a very small … **4** … in a 'big bang'. **4 marks**

All the remaining questions have four parts.
Choose only one answer in each part, but you can use the same number more than once.

Specimen QUESTION FIVE

Radioactive materials give off radiation.

5A Which row in the table is correct?

	made of particles?		
	alpha	beta	gamma
1	✔	✗	✗
2	✔	✗	✔
3	✗	✔	✔
4	✔	✔	✗

5B Which row in the table is correct?

	deflected by an electric field?		
	alpha	beta	gamma
1	✔	✔	✗
2	✔	✗	✗
3	✔	✗	✔
4	✗	✔	✔

5C A computer WiFi wireless network operates at 2.4 GHz (2 400 000 000 Hz).
If the wavelength is 0.125 m, what is the wave speed of the WiFi signal in m/s?

1 19 200 000

2 300 000 000

3 3 000 000 000

4 19 200 000 000

5D WiFi networks use digital signals rather than analogue signals.
Which statement is correct?

1 Digital signals are less prone to interference than analogue signals.

2 Analogue signals are more easily processed by computers.

3 Digital signals are continuously varying.

4 Digital signals can travel through optical fibres but analogue signals cannot. **4 marks**

Specimen QUESTION SIX

Half-life is the time needed for the number of nuclei of a radioactive isotope in a sample, to halve. Carbon-14 has a half-life of 5730 years. It is taken up by living things while they are alive. Objects made from animal or plant materials can be dated by finding the amount of carbon-14 they still contain.

6A A sample of ancient cloth contains 4 million carbon-14 atoms.
Which row in the table is correct?

	number of carbon-14 atoms in sample	
	5730 years ago	in 5730 years time
1	2 million	1 million
2	8 million	2 million
3	8 million	1 million
4	2 million	8 million

6B Beta radiation consist of …

1 helium nuclei released from carbon-14 nuclei.

2 high frequency electromagnetic radiation.

3 high energy electrons surrounding carbon-14 nuclei.

4 high energy electrons released from carbon-14 nuclei.

6C One reason why carbon-14 is not suitable as a tracer for medical images is that …

1 beta radiation is absorbed by the body before it reaches the detector.

2 beta radiation is the most ionising form of radiation.

3 only ancient materials contain carbon-14.

4 its half-life is too short.

6D Rocks may be millions of years old.
One reason why carbon-14 is not suitable for dating rocks is that …

1 rocks do not contain carbon.

2 it is difficult to prepare rock samples for dating.

3 beta radiation cannot escape from rocks.

4 the number of carbon-14 atoms in a rock would be too small to measure.

4 marks

Answers to exam questions

Unit B1a Human Biology

Question	A	B	C	D
ONE	2	1	4	3
TWO	3	1	4	2
THREE	2	1	4	3
FOUR	4	2	1	3
FIVE	4	2	1	1
SIX	1	1	2	4

Unit B1b Evolution and Environment

Question	A	B	C	D
ONE	2	4	3	1
TWO	1	4	3	2
THREE	3	2	4	1
FOUR	2	4	1	3
FIVE	3	4	3	4
SIX	2	3	4	1

Unit C1a Products from Rocks

Question	A	B	C	D
ONE	4	2	3	1
TWO	2	4	1	3
THREE	1	4	2	3
FOUR	3	4	1	2
FIVE	3	1	3	2
SIX	3	4	1	3

Unit C1b Oils, Earth and Atmosphere

Question	A	B	C	D
ONE	1	4	3	2
TWO	3	1	4	2
THREE	2	4	1	3
FOUR	4	3	1	2
FIVE	4	1	2	2
SIX	2	3	3	4

Unit P1a Energy and Electricity

Question	A	B	C	D
ONE	2	1	4	3
TWO	4	1	3	2
THREE	4	3	2	1
FOUR	3	1	4	2
FIVE	1	3	1	1
SIX	1	3	2	4

Unit P1b Radiation and the Universe

Question	A	B	C	D
ONE	1	4	2	3
TWO	3	2	4	1
THREE	4	3	1	2
FOUR	3	2	4	1
FIVE	4	1	3	1
SIX	2	4	1	4

Answers to practice questions

Biology B1a

Nervous system (p3)

1 It is the junction between two nerve cells.

2 hot object (stimulus) → receptor in skin → signal travels through a sensory neurone → signal travels through a relay neurone → signal travels through a motor neurone → muscle (effector) moves hand away

Hormones (p5)

1 Two from: water content, ion content, blood sugar level, temperature

2 Through the bloodstream

3 Allows both fast and slow responses and effects that are short-lived or long-lasting

Controlling reproduction (p7)

1 Pituitary gland. It causes eggs to mature and stimulates the ovaries to produce oestrogen.

2 Ovaries. It stimulates the pituitary gland to make LH and inhibits production of FSH.

3 FSH increases the woman's natural level of FSH and makes it more likely that eggs will mature. Oestrogen inhibits FSH production and can be used as an oral contraceptive.

Diet and exercise (p9)

1 A diet containing the correct nutrients in the right amounts, which supplies the right amount of energy

2 Problems such as: becoming too thin, periods in women stopping, reduced resistance to infection, named deficiency disease such as rickets from lack of vitamin D

3 Problems such as: becoming too fat, heart disease, high blood pressure, diabetes, arthritis

Cholesterol and salt (p11)

1 The liver

2 LDLs are 'bad' cholesterol and HDLs are 'good' cholesterol. LDLs take cholesterol from the liver to the cells and HDLs take cholesterol from the cells to the liver.

3 Processed foods often contain a lot of fat, which can increase body weight and cholesterol levels, leading to cardiovascular disease. Processed foods often contain a lot of salt, which causes high blood pressure in 30% of the population.

Medical drugs (p13)

1 A substance that affects the body

2 Laboratory testing with computer models and cells, then animal testing to look for side-effects, then testing in human volunteers

3 Thalidomide causes birth defects, so it cannot be prescribed if there is any chance of a woman being pregnant, or becoming pregnant, while they are taking it.

Recreational drugs (p15)

1 To become dependent on something and find it difficult to give up

2 Nicotine is the addictive substance, carbon monoxide reduces the oxygen-carrying capacity of the blood, tar causes cancer.

3 Nerves and brain, and, in the longer term, the liver

4 Smokers are addicted and will suffer unpleasant withdrawal symptoms if they try to give up. Social pressure from friends and family may make it difficult or undesirable to stop.

Pathogens and disease (p17)

1 A microorganism that can cause infectious disease

2 They reproduce rapidly inside the body and release toxins.

3 They ingest pathogens, they produce antitoxins, they produce antibodies that destroy particular pathogens.

Antibiotics (p19)

1 A substance that can kill bacteria

2 Colds are caused by viruses, not bacteria, and antibiotics only work against bacteria.

3 Antibiotic-resistant strains could appear if antibiotics are over-used.

Vaccination (p21)

1 Small amounts of dead or inactive pathogens

2 It causes the production of white blood cells that can produce antibodies against it. Some of these remain after the vaccine has been destroyed. If the pathogen infects the body, these white blood cells reproduce rapidly and produce a lot of antibodies very quickly.

3 Flu viruses mutate so one year's vaccine may not protect against the following year's strains of flu viruses.

Biology B1b

Adaptation (p23)

1 Three from: food, water, mates, territory

2 The tree competes with the grass for light, water and nutrients.

3 Camouflage against the snow in winter and the ground in summer

Reproduction (p25)

1 In the nucleus

2 The offspring get half their genetic information from one parent and half from the other, so the genetic information is mixed.

3 There is only one parent and no mixing of genetic information.

Cloning and genetic engineering (p27)

1 Taking cuttings and letting them grow. Tissue culture of small groups of cells

2 The host mother provides the womb for the embryo to grow in, but not its genetic information. The genetic information comes from the original embryo.

3 Use adult cell cloning: take the nucleus out of an egg cell, replace it with the nucleus from one of the horse's body cells, implant the egg into a host mother

Extinction and evolution (p29)

1 The remains of a prehistoric animal or plant, preserved in the Earth's crust

2 The rats might eat their eggs, or compete for food.

3 Simple life form developed on the Earth more than three billion years ago and all species of living things evolved from these.

Natural selection (p31)

1 Individuals in a species are not all the same. This is because of their genes. Individuals most suited to their environment survive to reproduce. Their offspring inherit the genes that made them successful.

2 It conflicted with religious views; people could not understand how complex features could evolve.

3 Bacteria are suited to their environment. They will continue to exist unless they cannot adapt to changes in their environment.

Pollution (p33)

1 The population and standard of living are increasing.

2 Chemicals such as fertilisers, herbicides and pesticides can be washed into rivers and lakes.

3 The air in towns is likely to be more polluted than the air in the countryside, so fewer lichens grow there.

4 Metal ores are a non-renewable resource, so they will run out one day. If we recycle metals, it will increase the time taken for metal ores to run out. They may not run out at all.

Global warming (p35)

1 A gas that is particularly good at absorbing and emitting thermal energy in the atmosphere

2 Any one of: rise in sea levels, big changes in climate, melting ice caps

3 Rice paddy fields and cattle release methane. Clearing forests for farming reduces the number of trees available to absorb carbon dioxide. Decay microorganisms and burning the trees release carbon dioxide.

Chemistry C1a

Atoms and elements (p37)

1 Central nucleus surrounded by electrons

2 A substance made from only one sort of atom

3 A column containing elements with similar properties

Reactions and compounds (p39)

1 $6 + 16 = 22\,g$

2 Each molecule contains six atoms of carbon, twelve atoms of hydrogen, and six atoms of oxygen.

3 They are moved from one atom to another, or shared between two atoms.

4 $C + 2CuO \rightarrow CO_2 + 2Cu$

Limestone (p41)

1 Advantages: local employment, produces useful raw material

Disadvantages: noisy, dusty, may reduce tourist trade, leaves a big hole

2 Large building projects need lots of concrete, steel (from iron), glass and new roads. Limestone is a raw material for producing these materials.

Ores and iron (p43)

1 A rock or mineral that contains enough metal to make it worthwhile extracting the metal

2 Lead is less reactive than carbon but aluminium is more reactive than carbon.

3 It is 96% iron. The impurities it contains makes it hard but brittle.

Alloys (p45)

1 A mixture of a metal with non-metals or other metals

2 In a pure metal the layers of atoms can slide over each other easily. In an alloy, the different-sized atoms distort the regular arrangement of atoms so that the layers do not slide over each other easily.

3 High carbon steel is hard. It would be difficult to press it into shape.

Copper, titanium and aluminium (p47)

1 It does not react with water and it is easily bent into shape.

2 A lot of energy is needed to extract them. Titanium is extracted in several steps. Aluminium is extracted using electricity.

3 Just one step should reduce the cost of extracting titanium. If titanium is cheaper, it should be used more widely.

Hydrocarbons (p49)

1 A compound of hydrogen and carbon only

2 C_6H_{14}

3 The crude oil is heated to evaporate it, then allowed to condense at different temperatures.

4 The fractions at the top contain shorter alkanes, which are runnier and easier to ignite.

Burning fuels (p51)

1 Five from: water vapour, carbon dioxide, carbon monoxide, particles, sulfur dioxide

2 Complete combustion happens when there is plenty of air. Incomplete combustion happens when there is not enough air. Carbon monoxide is produced instead of carbon dioxide in incomplete combustion.

3 Fewer particles will cause global dimming to decrease. More sunlight will reach the Earth's surface, increasing the amount of thermal energy that could be trapped by the atmosphere.

Chemistry C1b

Alkenes (p53)

1 Cracking is the breakdown of long alkanes to form shorter alkanes and alkenes. The alkanes are vaporised and passed over a hot catalyst. This is done because the demand for shorter alkanes cannot be met by fractional distillation alone.

2 C_6H_{12}

3 The ethene would come from crude oil and that would not sound very appealing.

Polymers (p55)

1 Poly(chloroethene)

2 HDPE would be better. HDPE is stronger than LDPE. Hot drinks would be too hot for LDPE but not for HDPE, which is useable up to 120 °C.

3 Alkenes used as monomers are obtained by cracking crude oil fractions. If the cost of oil goes up, so will the cost of polymers.

Plant oils (p57)

1 Crush the plant material. Press to remove the oil, or dissolve the oil in solvent then distil to remove the solvent.

2 They are a source of energy and nutrients such as vitamins.

3 A mixture of oil and water. Found in ice cream, salad dressings

4 React with hydrogen at about 60 °C in the presence of a nickel catalyst. Hardened vegetable oils are solid at room temperature and can be used in margarine, and for making cakes and pastries.

Food additives (p59)

1 Two from: improve taste, improve appearance, improve shelf life

2 The additives are listed on the ingredients label.

3 Banned food additives may get into food by mistake, some people have allergies to certain additives.

Structure of the Earth (p61)

1 It has a layered structure. The innermost layer is a core made from iron and nickel. The next layer is the mantle, made from rock that can flow very slowly. The outermost layer is a rocky crust.

2 Mountains should be all over the surface and they are not.

3 The coasts of South America and Africa fit together; similar rocks are found on both continents; similar fossils are found on both continents.

4 Boundaries between tectonic plates

The atmosphere (p63)

1 Carbon dioxide

2 Nitrogen

3 Carbon dioxide: dissolved in oceans, formed fossil fuels, formed carbonate rocks

Oxygen: produced by plants because of photosynthesis

4 These events happened a very long time ago. No-one was there to see them and record observations, so the theories must be based on evidence left behind.

Noble gases (p65)

1 Three from: helium, neon, argon, krypton, xenon, radon

2 It is much less dense than air.

3 It is unreactive, so it stops the metal filament burning away.

4 The noble gases are unreactive, so do not form compounds.

Physics P1a

Heat transfer (p67)

1 The thermal energy travels through space as infra red radiation.

2 The lid reduces heat loss by convection.

3 Shiny surfaces are poor absorbers and emitters of thermal energy. The shiny foil helps to stop the satellite or spacecraft getting too hot or too cold.

Efficiency (p69)

1 Energy that is not usefully transferred or transformed

2 The surroundings warm up and the energy becomes spread out too thinly to be useful.

3 35 MJ

4 0.7

Cost of electricity (p71)

1 The watt, W

2 energy transferred = power × time

3 $250 \div 1000 \times 8 = 2$ kWh

4 $100 \div 1000 \times 6 \times 10 = 6$p

The National Grid (p73)

1 The network of cables that transfers electricity from the power stations to the consumers

2 High voltages reduce the current in the cables, which reduces energy losses from them.

3 Step-up transformers increase the voltage from the power stations for transmission in the cables. Step-down transformers reduce the voltage to safer levels for distribution to consumers.

Generating electricity (p75)

1 Fossil fuels: oil, coal, natural gas. Nuclear fuels: uranium, plutonium

2 Thermal energy released from the fuel is used to boil water. The steam drives a turbine. This is connected to a generator, which turns and produces electricity.

3 There are several energy transfer and transformation steps, and each of these will produce waste energy.

Renewable energy sources (p77)

1 Wave machines use the rise and fall of water because of waves. Tidal barrages use the kinetic energy in moving water as the tide changes. Hydroelectric power schemes use the kinetic energy from water stored in dams to turn electricity generators.

2 Satellites are too high up to run cables to them. There is no night or cloud in space, and the cells should supply electricity constantly unless the satellite goes into the Earth's shadow.

3 If the thermal energy is extracted too quickly, the rocks will become too cold to be useful.

Resources compared (p79)

1 They can be started up quickly but their fuel is expensive.

2 They take a long time to start up but their fuel is cheap.

3 The advantages of each type of power station can be exploited and their disadvantages overcome.

Physics P1b

The electromagnetic spectrum (p81)

1 wave speed = frequency × wavelength

2 wave speed = $2\,400\,000\,000 \times 0.125 = 300\,000\,000$ m/s

3 Ultraviolet

Some uses of electromagnetic radiation (p83)

1 We feel it as heat.

2 Water, fats and sugars in food easily absorb microwaves. The molecules vibrate more vigorously and the food heats up.

3 The X-rays pass through the metal parts allowing a picture to be taken without taking the machinery apart.

Communication signals (p85)

1 Analogue signals are continuously varying but digital signals have certain states, usually 0 and 1. Digital signals are affected less by noise.

2 The light is repeatedly reflected inside the fibre.

3 Microwaves pass through the atmosphere.

Hazards of electromagnetic radiation (p87)

1 X-rays are absorbed by lead.

2 The clothing absorbs ultraviolet light and reduces exposure to it.

3 Metal mesh in the door to stop microwave radiation escaping and safety lock so the oven does not work with an open door.

Atomic radiation (p89)

1 Atoms that have the same number of protons but different numbers of neutrons

2 Alpha radiation and beta radiation

3 Two from: alpha is stopped by paper but beta and gamma are not; alpha and beta are stopped by thin aluminium but gamma is not; gamma travels further in air than alpha and beta; alpha is the most ionising and gamma is the least ionising

Half-life (p91)

1 One from: the time taken for half the nuclei of that isotope in a sample to decay; or the time taken for the count rate from a sample to fall to half the starting rate

2 0.5 g ($240 \div 60 = 4$ half-lives, so $8\,g \rightarrow 4\,g \rightarrow 2\,g \rightarrow 1\,g \rightarrow 0.5\,g$)

3 It is a gamma emitter so it passes out of the patient's body without harming them and is easily detected, and it has a short half-life so that it will not persist in the body for a long time.

Observing the universe (p93)

1 They work during the day and in bad weather; they provide more detailed images.

2 The idea that the universe began from a very small point about 13.6 billion years ago, and expanded rapidly. It continues to expand.

3 The light from galaxies is red-shifted, showing that the galaxies are moving away from us. The more distant the galaxy, the more its light is red-shifted. This shows that more distant galaxies are moving more quickly. These features are shared by explosions, where the fastest objects travel furthest.

A

adaptation Feature of an organism that makes it suited for living in its habitat

alloy A mixture of a metal with non-metals or other metals

alpha radiation Atomic radiation consisting of alpha particles, two protons and two neutrons joined together.

amplitude The distance between the maximum disturbance of a wave and its undisturbed position

analogue signal Communication signal that varies continuously

antibiotic Chemical that can kill bacteria

antibody Protein produced by white blood cells, capable of attaching to particular antigens

antigen Substance recognised as foreign to the body

antitoxin Chemical produced by white blood cells that counteracts a toxin

asexual reproduction Producing offspring from just one parent

atmosphere Layer of gas surrounding the Earth

atom Tiny particle that all substances are made from

B

bacterium (plural **bacteria**) Tiny single-celled organism

balanced diet A diet that contains the required nutrients in the correct amounts for health

balanced equation Description of a chemical reaction that shows the number and formulae of each substance involved

beta radiation Atomic radiation consisting of beta particles, electrons from the nucleus

Big Bang theory The idea that the universe began as a very small point that expanded rapidly, and continues to do so

biodegradable Able to be broken down by microorganisms

bond Join or force of attraction between two atoms

C

catalyst A substance that speeds up chemical reactions without being used up itself

chemical bond See *bond*

chemical formula Combination of numbers and chemical symbols that shows how many of each sort of atom are in a compound

chemical symbol One, two or three letters that represent an element

cholesterol A substance made using saturated fats in the liver

chromatography Method used to detect and identify coloured substances

chromosome Object in the nucleus carrying genes

clone Genetically identical individual

combustion Burning – the reaction of a fuel with oxygen

compound Substance made from more than one element chemically joined together

condense To turn from a gas to liquid

conduction The transfer of thermal energy by passing vibrations from one particle to the next in a substance

convection The transfer of thermal energy in a liquid or gas by particles moving from place to place

convection current The circulating movement of liquids or gases caused by heating

core Innermost layer of the Earth, made from iron and nickel

corrosion The reaction of a metal with air and water, such as rusting

cracking Thermal decomposition reaction in which longer alkanes are broken down into shorter alkanes and alkenes

crust Outermost layer of the Earth, made from rock

D

density The mass of a piece of material divided by its volume

digital signal Communication signal with discrete states: 0 and 1, or ON and OFF

drug A substance that changes chemical processes in the body

E

E number Unique number given to food additives to show that they have been tested and passed as safe to use

effector Part of the body that responds to a stimulus, such as a muscle or gland

efficiency A measure of the proportion of energy supplied to a device that is usefully transferred

electron Particles in an atom arranged around its nucleus

element Substance from just one type of atom

embryo transplant Separating cells from an early embryo and transplanting the identical embryos into host mothers

emulsion Mixture of oil and water

environment The conditions in a habitat

ethanol The alcohol in alcoholic drinks, C_2H_5OH

evaporate To turn from a liquid to a gas

evolution Development of a type of living thing from an earlier, simpler life form

extinct When a species no longer has any living members left

F

fertilisation Joining or fusion of gametes

food additive Natural or artificial substance added to food to improve its properties

fossil fuel Fuel formed from the ancient remains of dead plants or animals

fractional distillation Method used to separate mixtures which relies on the different substances having different boiling points

fractionating column The equipment used to separate crude oil by fractional distillation

fractions The individual parts separated by fractional distillation

frequency The number of waves that pass a certain point each second, measured in hertz, Hz

FSH Hormone secreted by the pituitary gland which causes eggs to mature

G

gametes Sex cells. Sperm are male sex cells and eggs are female sex cells.

gamma radiation Part of the electromagnetic spectrum, used for sterilising surgical instruments

gene Section of DNA in a chromosome that controls a characteristic in the body

generator Machine that transforms kinetic energy into electrical energy

genetic engineering Transferring genes from one cell to another

genetic modification Another name for genetic engineering, shortened to GM

geothermal energy Thermal energy from hot rocks deep underground

gland Tissue or organ able to produce hormones

global dimming Reduction in sunlight reaching the Earth's surface because of particles produced by burning fuels

global warming Worldwide increase in average temperatures

greenhouse effect Absorption of thermal energy in the atmosphere

greenhouse gas Atmospheric gas that is particularly good at absorbing and emitting thermal energy

group Column of elements in the periodic table that have similar properties

group 0 The noble gases, found in the far right hand column of the periodic table

H

habitat The place where an organism lives

half-life The time taken for half the nuclei of the isotope in a sample to decay; or the time taken for the count rate from a sample to fall to half the starting rate

hardened Vegetable oil reacted with hydrogen

HDL High-density lipoprotein, also known as 'good' cholesterol

hormone Chemical produced by a gland and transported in the bloodstream to a target organ

hydrocarbons Compounds containing hydrogen and carbon only

hydroelectric power scheme A scheme to generate electricity from water held behind a dam

I

immune Protected against a particular infective pathogen and the disease it causes

infra red radiation Thermal radiation, also called infra red light. It is part of the electromagnetic spectrum and is used for TV remote controls.

insulator A poor conductor of thermal energy

ion Charged particle made when an atom loses (or gains) an electron

ionising radiation Radiation that produces charged particles from atoms

isotopes Atoms with the same number of protons but different numbers of neutrons

IVF *In vitro* fertilisation – fertilising an egg outside the body in a laboratory

J

joule The unit of energy, symbol J

K

kilowatt-hour kWh – the unit of electrical energy transferred from the mains supply

L

LDL Low-density lipoprotein, also known as 'bad' cholesterol

limestone A rock made from calcium carbonate, $CaCO_3$

lipoprotein A combination of cholesterol and protein found in the blood

M

mantle Layer of the Earth between the crust and core, made from rock that can flow very slowly

menstrual cycle Monthly series of changes in the reproductive system of women

metabolic rate The speed at which the chemical reactions in the body's cells happen

microwaves Part of the electromagnetic spectrum, used for communications and cooking food

molecular formula Formula showing the number of atoms of each element in a compound

monomers Small molecules that can join together to make longer molecules

motor neurone Nerve cell that carries information to an effector

N

National Grid The network of cables that transfers electricity from power stations to consumers

natural selection The theory of how species change over time

neurone Nerve cell

neutron Neutral particle in the nucleus

nicotine The addictive substance in tobacco smoke

nuclear fission Change to an atom in which its nucleus breaks apart

nuclear reaction Change to the nucleus of an atom

O

oestrogen Hormone secreted by the ovaries which inhibits production of FSH

optical fibre Thin rod of glass that can carry infra red and visible light signals

ore A rock or mineral that contains enough metal to make it worthwhile extracting the metal

P

pathogen Microorganism that causes infectious disease

periodic table Chart showing the elements

pollutant Substance produced that harms the environment

pollution Waste products that harm the environment when released

polymer A large molecule formed from many monomers by polymerisation

power The rate of energy transfer

product A substance that is made in a chemical reaction

proton Positively charged particle in the nucleus

Q

quicklime Calcium oxide, CaO

R

radiation invisible particles or rays given off by nuclei of atoms

radio waves Part of the electromagnetic spectrum, used for transmitting television and radio programmes

radioactive Able to give off radiation

reactant A substance that reacts in a chemical reaction

receptor Group of cells that detects a stimulus

red-shift The observed light from distant galaxies is moved towards the red end of the spectrum.

reduction Reaction in which oxygen is removed from a substance

relay neurone Nerve cell that passes information from a sensory neurone to a motor neurone

renewable energy source A source of energy that will not run out provided it is managed effectively

renewable resources Substances that will not run out

S

saturated Contains only single bonds

sensory neurone Nerve cell that carries information from a receptor

sexual reproduction Producing offspring from two parents

slaked lime Calcium hydroxide, $Ca(OH)_2$

smart alloy Mixture of metals that can return to its original shape after being bent

solar cell A device that converts light energy directly into electrical energy

stimulus Change in the surroundings detected by a receptor

structural formula Diagram showing how the atoms in a compound are joined together

sustainable development Living in such a way that we meet our needs without damaging the ability of future generations to meet their own needs

synapse Junction between two nerve cells

T

tectonic plate Piece of the crust and upper mantle

theory of continental drift The idea that tectonic plates move, carrying continents with them

thermal decomposition The chemical reaction where a compound breaks down into simpler substances when heated

thermal energy Heat energy

tidal barrage A barrier built across a river mouth to take advantage of moving water produced by the tides

tobacco Dried leaves from the tobacco plant, used in cigarettes

toxin Poison

tracer Radioactive substance used to follow chemicals in the environment or in the body

transformer A device that changes the voltage of an electrical supply

turbine Machine with many blades. It spins when steam is passed into it.

U

ultraviolet radiation Part of the electromagnetic spectrum, used for sun beds

unsaturated Containing a carbon–carbon double bond

V

vaccination The process of providing a vaccine to make someone immune to a particular pathogen

vaccine A small amount of dead or inactive pathogens, usually injected into the body

virus Tiny particle of genetic information surrounded by a protein coat

viscosity A measure of how runny a liquid is – water has a low viscosity and treacle has a high viscosity

volatile Liquids that easily turn into gases are very volatile.

W

watt The unit of power, symbol W

wave machine Device that converts the kinetic energy in waves into electrical energy

wavelength The distance between a point on one wave and the same point on the next wave, measured in metres, m

white blood cell A type of cell found in the blood, and which helps to defend the body against pathogens

wind turbine Type of windmill that generates electricity using the kinetic energy in wind

X

X-rays Part of the electromagnetic spectrum, used for making medical images of broken bones

Last-minute learner

Biology Unit B1a Human Biology

Controlling body processes

- Nerve cells are neurones. The gap between two neurones is a synapse. Chemicals diffuse across the gap so that signals can pass from one neurone to another.
- Reflex actions are fast and automatic.

part of the nervous system	what it does
receptor cell, e.g. in the skin	detects a stimulus, such as a change in temperature
sensory neurone	carries information from a receptor
relay neurone in spinal cord	carries information from a sensory neurone to a motor neurone
motor neurone	carries information to an effector
effector, e.g. a muscle or gland	carries out a response, such as a muscle moving or a gland producing a hormone

- Internal conditions in the body must be controlled, including water and ion content, blood sugar level and temperature.
- Water and ions leave the body from the skin and kidneys. Water also leaves the body from the lungs.
- Hormones are chemicals produced by glands. They are transported in the bloodstream to their target organs, where they cause a change.
- Hormones control the menstrual cycle in women, which lasts for about 28 days.

hormone	produced by	what it does
FSH	pituitary gland	causes an egg to mature stimulates oestrogen release
oestrogen	ovaries	stops FSH production simulates LH release causes the womb lining to thicken
LH	pituitary gland	causes the release of an egg at about day 14

- Oral contraceptives contain hormones that stop production of FSH, so eggs do not mature.
- Fertility drugs boost the levels of FSH. This helps more eggs mature for use in *in vitro* fertilisation or IVF.

Diet and health

- People become too thin if they do not eat enough. They become fat if they eat too much or take too little exercise.
- Lack of food can cause health problems, including irregular periods in women and reduced resistance to infection.
- The metabolic rate is the speed of chemical reactions in the body. Active people with a lot of muscle and little fat have high metabolic rates. Exercise and inherited factors also influence it.
- Heart disease, high blood pressure, arthritis and diabetes are linked to excess weight.
- Cholesterol is made by the liver from saturated fats.
- Low-density lipoproteins or LDLs are 'bad' cholesterol. They carry cholesterol from the liver to the body's cells.
- High-density lipoproteins or HDLs are 'good' cholesterol. They carry cholesterol from the body's cells to the liver.
- Processed foods often contain a lot of fat and salt. Too much salt causes high blood pressure, leading to cardiovascular disease.

Drugs

- Drugs change chemical processes inside the body. Many drugs are found naturally. New drugs are tested to make sure they are effective and safe.
- Thalidomide caused birth defects during the early 1960s. It is now used to treat leprosy, but cannot be given to pregnant women.
- Alcohol slows down reactions and leads to a loss of self-control. It can damage the liver and brain.

substance in tobacco smoke	effect on body
nicotine	addictive
carbon monoxide	reduces the ability of the blood to carry oxygen
tar	causes cancer

Pathogens and disease

- Pathogens are micro-organisms, such as bacteria and viruses, that cause disease.
- Bacteria multiply rapidly in the body and release toxins. Viruses multiply inside cells and damage them when they escape.
- White blood cells produce antitoxins and antibodies, and they ingest pathogens. Antibodies stick to certain antigens carried by pathogens.
- Antibiotics kill bacteria but not viruses. Over-use of antibiotics has produced resistant strains of bacteria because of natural selection.
- Vaccination involves injecting a dead or inactive form of a pathogen. White blood cells are produced that recognise the pathogen and produce antibodies against it.

Biology Unit B1a Evolution and Environment

Adaptation

- Plants compete with each other for light, water, nutrients and space.
- Animals compete with each other for food, water, mates and territory.
- Organisms have particular features, called adaptations, which allow them to survive in their environment.
- Extinction may be caused by changes to the environment, new competitors, new diseases and new predators.
- The fossil record provides evidence of how life on Earth has changed over millions of years.

Evolution

- The theory of evolution states that life developed on Earth more than three billion years ago, and that all species have evolved from these early life forms.
- Charles Darwin proposed the theory of evolution by natural selection. Individuals in a species show variation caused by differences in their genes. Individuals with features best suited to the environment are more likely to survive to reproduce and pass on their beneficial features.
- Darwin's theory was only gradually accepted and still causes controversy.
- Other theories of evolution have been proposed, but they have been disproved.

Reproduction

- Chromosomes are found in the cell's nucleus. They contain many genes, each of which controls the development of a characteristic in an organism.
- Asexual reproduction involves just one parent. All the offspring are identical because there is no mixing of genetic information. Genetically identical individuals are called clones.
- Sexual reproduction involves two parents. Fertilisation is when a male gamete and a female gamete fuse together. Genetic information is mixed, leading to variety in the offspring.

Cloning and genetic engineering

- Plants can be cloned by taking cuttings and by tissue culture.
- Animals can be cloned by embryo transplants. A developing embryo is split into separate cells at an early stage, grown in the laboratory and then transplanted into host mothers.
- Animals can also be cloned by removing the nucleus from an unfertilised egg and replacing it with the nucleus from a different cell. It is called fusion cell cloning if the nucleus comes from an embryo, and adult cell cloning if it comes from a body cell of an adult animal.
- Genetic engineering is also called genetic modification or GM. Genes are moved from one organism to another using enzymes.

Pollution

- The human population is increasing rapidly and the standard of living is improving. This increases the use of non-renewable energy resources and raw materials, and increases the amount of waste.
- Sustainable development involves meeting our needs without damaging the ability of future generations to meet their own needs.
- Waste causes pollution if it is not handled properly.
- The air can be polluted with gases such as sulfur dioxide, which causes acid rain.
- The land can be polluted with toxic chemicals, such as pesticides.
- Rivers, lakes and seas can be polluted with toxic factory waste, sewage, and chemicals washed off farmland.
- Lichens are plants that can be used as indicators of air pollution. Insect larvae and other invertebrates can be used as indicators of water pollution.

Global warming

- Carbon dioxide and methane are greenhouse gases. They absorb thermal energy and keep the planet warmer than it would be without them.
- Increasing levels of greenhouse gases leads to global warming. Increases of average temperatures by only a few degrees may cause a rise in sea levels and climate change.
- Extra methane is produced by cattle and rice paddy fields.
- Extra carbon dioxide is produced by burning fossil fuels. Deforestation makes the situation worse because there are fewer trees to absorb the carbon dioxide.

Chemistry Unit C1a Products from Rocks

Atoms and elements
- Atoms have a nucleus at the centre with electrons arranged around it.
- An element is made of just one sort of atom. There are about 100 different elements, each with a unique chemical symbol.
- The periodic table is a chart showing information about the elements.
- Similar elements are found in columns of the periodic table called groups.

Reactions and compounds
- Reactants are the chemicals that react together, and products are the chemicals made.
- Chemical bonds between atoms involve sharing or transferring electrons between atoms.
- A compound is made of the atoms of two or more elements bonded together. A chemical formula shows the number and type of each atom present.
- No atoms are lost or made in a chemical reaction, so the total mass stays the same. Balanced equations describe reactions.

Limestone
- Limestone contains calcium carbonate, $CaCO_3$. It is used as a building material, and in the manufacture of glass, iron, cement and concrete.
- Calcium carbonate breaks down when heated to form calcium oxide (CaO) and carbon dioxide. This is called thermal decomposition. Other carbonates react in a similar way. Calcium oxide is also called quicklime.
- Calcium oxide reacts with water to form calcium hydroxide, $Ca(OH)_2$. This is also called slaked lime.

Ores and iron
- An ore is a rock that contains enough metal to make it economical to extract the metal.
- Unreactive metals are found as the metals themselves.
- Most metals are found as compounds. Metal oxides are reduced to extract the metal. If the metal is less reactive than carbon, it is extracted using carbon or carbon monoxide. Iron is extracted this way.
- Iron from the blast furnace is too brittle for most uses and pure iron is too soft. Iron is usually converted into steel.

Alloys
- An alloy is a mixture of a metal with other metals or non-metals. Steel is iron mixed with metals such as vanadium.
- Alloys are harder than individual metals because they have different sized atoms. This stops layers of atoms sliding over each other easily.
- Smart alloys have unusual properties. Shape memory alloys can be twisted and bent, but snap back into shape when warmed up.

Copper, titanium and aluminium
- Transition metals are the elements in the central block of the periodic table, such as iron, copper and titanium.
- Most high-grade copper ores have already been mined. Low-grade ores produce a lot of waste. Copper is purified by passing electricity through its solutions. Bacteria are also used to extract copper.
- Titanium cannot be extracted using carbon. Several steps are needed. Each one needs a lot of energy.
- Aluminium is in group 3. It is more reactive than carbon so it is extracted using electricity.

Alkanes
- Crude oil contains hydrocarbons, compounds of hydrogen and carbon only.
- Alkanes are unsaturated, which means that they only have single bonds. Their general formula is C_nH_{2n+2}. For example, ethane is C_2H_6.
- Fractional distillation separates the different alkanes in crude oil. This works because they have different boiling points.
- In fractional distillation, oil is heated to evaporate the hydrocarbons. The vapours rise up a tower, which is hotter at the bottom. Different fractions condense at different temperatures.

Burning fuels
- Hydrocarbons burn in plenty of air to form water vapour and carbon dioxide. If there is not enough air, carbon monoxide and particles of carbon are produced instead of carbon dioxide.
- Carbon dioxide is a greenhouse gas. Carbon particles cause global dimming.
- Many fuels contain sulfur. They produce sulfur dioxide when they burn, which causes acid rain.

Chemistry Unit C1b Oils, Earth and Atmosphere

Alkenes

- Alkenes are unsaturated hydrocarbons because they contain a carbon–carbon double bond. Their general formula is C_nH_{2n}. For example, ethene is C_2H_4.
- Alkenes decolourise brown bromine water but alkanes do not.
- Alkenes are produced by heating alkanes over a hot catalyst. This is called cracking. Cracking also produces shorter alkanes, which helps to meet the demand for fuels.

Products from alkenes

- Ethene reacts with steam at high temperatures and pressures to form ethanol. Ethanol can also be produced by fermentation of sugar.
- Alkenes can join together to make long molecules called polymers. The alkenes are called monomers when they do this.
- The properties of polymers depend on their monomers and the conditions used to join them together. Plasticisers are added to make polymers more flexible.
- Many polymers are not biodegradable. They do not rot away easily. This makes them difficult to dispose of after use.

Plant oils

- Vegetable oils are found in fruit, seeds and nuts. They are extracted by crushing plant materials and squeezing the oil out. They can also be extracted from crushed plant material using solvents.
- Vegetable oils are part of a balanced diet. They provide energy and contain certain vitamins. They are also used as fuels such as biodiesel.
- An emulsion is a mixture of oil and water. Emulsions are more viscous than oil or water. Emulsifiers are added to stop the oil and water separating. Emulsions are used in foods such as ice cream.
- Hydrogenated vegetable oils are used in margarine. Vegetable oils are hardened by reacting them with hydrogen.

Food additives

- Food additives are put in processed foods to improve the taste, appearance and shelf-life.
- Additives with an E number have been tested and passed as safe.
- Some food additives can cause allergic reactions in some people.
- Chromatography is used to detect and analyse artificial food colourings.

Structure of the Earth

- The Earth has a layered structure consisting of the core, mantle and crust.
- The upper part of the mantle and crust are broken into tectonic plates. These move at a few centimetres per year because of convection currents in the mantle, driven by heat from radioactive processes inside the Earth.
- The crust becomes unstable where tectonic plates meet. Earthquakes and volcanic eruptions happen there, but these are difficult to predict.
- At one time people thought that the Earth's surface features were caused by the crust shrinking as the Earth cooled.
- Alfred Wegener proposed the theory of continental drift. He could not explain how tectonic plates could move. It took many years for his theory to be generally accepted by scientists.

The atmosphere

- The Earth's atmosphere is about 80% nitrogen and 20% oxygen. There are small proportions of other gases such as carbon dioxide, water vapour and argon.
- Argon is a noble gas found in group 0. These gases are very unreactive. They are used in electric discharge tubes. Helium is a noble gas used in party balloons and airships.
- The Earth's atmosphere has been much the same for the last 200 million years.
- Scientists believe that the early atmosphere was mostly carbon dioxide, like the atmospheres of Mars and Venus today. When plants evolved, they absorbed carbon dioxide and released oxygen because of photosynthesis. Carbon dioxide also dissolved in the oceans, and became locked up in fossil fuels and carbonate rocks such as limestone.

Physics Unit P1a Energy and Electricity

Heat transfer
- Heat energy is also called thermal energy. The movement of atoms and molecules increases when they absorb thermal energy.
- Thermal energy can be transferred by conduction, convection and infra red radiation. Conduction and convection need particles, but radiation does not.
- Dark, dull surfaces are good emitters and absorbers of infra red radiation. Light, shiny surfaces are poor emitters and absorbers.
- The rate of transfer of thermal energy depends on the type of material, size of and shape of the object, and the difference in temperature.

Efficiency
- Energy is measured in joules, J. Energy cannot be created or destroyed, only transferred from place to place or transformed from one type to another.
- Energy that is not usefully transferred or transformed is 'wasted' energy.
- The efficiency of a device =
$$\frac{\text{useful energy transferred by device}}{\text{total energy supplied to device}}$$

Cost of electricity
- Power is measured in watts, W.
- Electrical energy from the mains supply is measured in kWh. 1 kWh is 1 unit of electricity.
- Electrical energy transferred (kWh) = power (kW) × time (h)
- Total cost of electricity = number of kWh × cost per kWh

The National Grid
- The National Grid transfers mains electricity from power stations to consumers.
- Thermal energy is lost from electricity cables. Electricity is transmitted at high voltages so that the current is low.
  The lower the current, the lower the amount of thermal energy lost.
- Step-up transformers increase the voltage from power stations so that energy losses are reduced. Step-down transformers decrease the voltage from the cables to safer levels for supply to homes.

Power stations
- Electricity is generated in power stations.
- Most electricity is generated in nuclear power stations or power stations fuelled by fossil fuels.
- Uranium and plutonium are nuclear fuels. Coal, oil and natural gas are fossil fuels. These fuels are non-renewable: they cannot be replaced once they have been used up.
- Thermal energy released from the fuel is used to boil water. The steam turns a turbine which drives an electricity generator.

Renewable energy sources
- Renewable energy sources will not run out, as long as they are managed effectively. There are no fuel costs but the machinery needed to use them can be expensive to build.
- Wind turbines use kinetic energy from the wind to drive electricity generators. They only work when the wind speed is high enough.
- Wave machines use the kinetic energy in waves to drive electricity generators. It has proved difficult to design machines that produce large amounts of electricity.
- Tidal barrages drive electricity generators using the moving water as the tides turn. Hydroelectric power schemes use moving water from behind dams to drive electricity generators.
- Solar cells convert light energy directly into electrical energy. They only work in the light. Geothermal power stations use hot water and steam produced deep inside the Earth.

Physics Unit P1b Radiation and the Universe

Waves
- Waves move energy from one place to another.
- The wavelength of a wave is the distance between a point on one wave and the same point of the next wave.
- The frequency of a wave is the number of waves per second, measured in hertz, Hz.
- wave speed (m/s) = frequency (Hz) × wavelength (m)

The electromagnetic spectrum
- Electromagnetic radiation travels as waves. All electromagnetic radiation travels at the same speed in a vacuum, 300 000 000 m/s.
- The main types of electromagnetic radiation, in order of increasing wavelength and decreasing frequency are: gamma rays, X-rays, ultraviolet rays, visible light, infra red rays, microwaves and radio waves.
- Gamma rays are used to sterilise surgical instruments.
  X-rays are used to make medical images of bones. Ultraviolet rays are used in sun beds. Microwaves and radio waves are used for communications.

Communication signals
- Communication signals may be analogue or digital.
- Analogue signals vary continuously and digital signals have discrete states, usually 0 and 1.
- Digital signals are less affected by noise than analogue signals.
- Light travels through optical fibres, even when the fibre is bent.

Atomic radiation
- Atoms consist of a central nucleus surrounded by electrons. The nucleus contains protons and neutrons.
- Isotopes are atoms with the same number of protons but different numbers of neutrons.

- Radioactive substances give out radiation from the nuclei of their atoms. Radioactive isotopes are called radioisotopes.
- Alpha radiation, α, consists of alpha particles. An alpha particle contains two protons and two neutrons. It is identical to a helium nucleus.
- Beta radiation, β, consists of high-energy electrons emitted from the nucleus.
- Gamma radiation, γ, consists of electromagnetic radiation emitted from the nucleus.
- Alpha radiation is stopped by paper, beta radiation by thin aluminium sheeting, and gamma radiation by thick lead.
- Alpha radiation and beta radiation are deflected by electric and magnetic fields, but gamma radiation is not.
- The half-life of an isotope is the time taken for half of the isotope in a sample to decay, or the time taken for the count rate in a sample to fall to half the starting rate.

Observing the universe
- Telescopes observe the universe at different frequencies of electromagnetic radiation. Ground-based optical telescopes only work at night in clear skies but radio telescopes work all the time. Space telescopes also work all the time, but are expensive and difficult to maintain.
- The Big Bang theory of the universe states that the universe began as a tiny single point that rapidly enlarged. The universe is still expanding.
- The wavelength and frequency of a wave source changes if the source moves towards or away from you.
- Light from distant galaxies is red-shifted. This shows that the galaxies are moving away from us.
- The light from the most distant galaxies is the most red-shifted. This shows that the most distant galaxies are moving away fastest.
- Red-shift evidence supports the Big Bang theory.